The New Alchemists

WITPRESS

WIT Press publishes leading books in Science and Technology.
Visit our website for the current list of titles.
www.witpress.com

WITeLibrary

Home of the Transactions of the Wessex Institute, the WIT electronic-library
provides the international scientific community with immediate and permanent
access to individual papers presented at WIT conferences.
Visit the WIT eLibrary athttp://library.witpress.com

The New Alchemists

the risks of genetic modification

M. Bizzarri
Roma University La Sapienza, Italy

WITPRESS Southampton, Boston

M. Bizzarri

Roma University La Sapienza, Italy

Published by

WIT Press

Ashurst Lodge, Ashurst, Southampton, SO40 7AA, UK
Tel: 44 (0) 238 029 3223; Fax: 44 (0) 238 029 2853
E-Mail: witpress@witpress.com
http://www.witpress.com

For USA, Canada and Mexico

WIT Press

25 Bridge Street, Billerica, MA 01821, USA
Tel: 978 667 5841; Fax: 978 667 7582
E-Mail: infousa@witpress.com
http://www.witpress.com

British Library Cataloguing-in-Publication Data

A Catalogue record for this book is available
from the British Library

ISBN: 978-1-84564-662-2
eISBN: 978-1-84564-663-9

Library of Congress Catalog Card Number: 2012932241

Contents

Foreword

[...] This is what boggles the mind here: "We can send a man to the moon but we can't stop a carp from reaching the Great Lakes?" said Bill Schuette, the attorney general of Michigan, which has led a legal and political fight to close locks that allow water to flow between the Mississippi River and the Great Lakes and, ultimately, to separate those two water systems entirely.

What to Do About Asian Carp? Great Lakes States Can't Agree.
The New York Times, 21 December 2011

Barring some catastrophic celestial event, Halley's comet will be passing by our planet Earth in mid-2061 and again about every 75–76 years later. Despite its temporal variability, this celestial event repeats a pattern that has been recorded since 240 BC. The variability in the dates has been accounted by equally valid demonstrations that there are interactions between planets and the comet that affect the latter's orbit. However, Halley's comet will show up pretty much when it is supposed to in this part of the Universe. In other words, the structure of determination of classical physical events is well understood. Instead, in the biological world in which all we humans participate, I cannot be so sure that I'll be able to finish typing the sentence that I am in the process of constructing (despite odds against it, I just did). This extreme distinction between the physical and biological worlds is based on the not well-appreciated fact that the biological world is mostly unpredictable. In other words, in the world we live in, constant change makes all recent and past events "original", in the sense that everything that happens or is about to happen is new and unpredictable. No two biological events are absolutely identical.

To reiterate it and make it more personal, you, the reader of this Foreword, are not the same person who started reading the current paragraph. You've changed in the process, not only because you are older by a few seconds but because you cannot ever go back to what you were, since many things in your environment have changed since you started reading this sentence. That's it ... in life, nothing is repetitively equal and everything is new. Of course, the changes I'm talking about are small, because it has only been seconds since you last checked whether a new wrinkle has appeared on your face, or whether a few new plaques have formed in your arteries. Regardless of these seemingly inconsequential factoids, you are not the same person who started reading this paragraph. You, and all of us, have simultaneously aged in many ways. Sparing you, the reader, much detail, this is one of the great differences between the worlds of classical physics and biology.

Nevertheless, it is not surprising to read, in the above epigram, a concerned lawyer failing to appreciate the difference between a mechanical event, designed and calculated on well-tested physical laws (lifting a significant payload and depositing it on the Moon's surface), and a biological phenomenon that was equally well-intended but badly designed and conceived and not tested before it was implemented. Presumably, the presence of Asian carp in the Mississippi river occurred despite careful planning by well-meaning biologists, and by social and environmental specialists who could not anticipate the consequences of human errors in dealing with continuous changes in complex niches like those in an enormous natural waterway.

The consequences of failures in either of the two events mentioned in the newspaper clipping (launching of a manned lunar-bound capsule and the introduction of Asian carp into the USA in the 1970s to control algal proliferation in aquaculture and wastewater treatment facilities) are quite significant for any fair-minded observer. In the case of space travel, less than a dozen highly intelligent volunteers accidentally died as a result of now well-known technical mistakes that were pointed out *a posteriori*. Once those design mistakes were corrected, fatal accidents were prevented. In any case, no one wishes to see volunteers die as a result of negligence or incompetence in applying physical laws. To the contrary, in the case of the imported Asian carp, the damage is counted in billions with no end of the story in sight. It would be pointless to list the unanswered questions regarding this biological event. One may even safely say that there will be no biological solution of the present problem. At most, there will be some accommodation, whereby the environment may take care of the changes caused by human intervention, and there will be some "new" bad and perhaps even some "new" good outcomes depending on who judges and when the evaluations are made. But for the most part, factually, we are unable to predict what will happen and which remedial measure will make the least harmful difference in the short run, let alone in the long run. In *The New Alchemists – The Risks of Genetic Modification*, Mariano Bizzarri expresses the need to consider the precautionary principle to prevent unpredictable damage caused by tampering with organisms. Unfortunately, as this book documents, there have been many instances when men played god, and I'm afraid that we have yet to see the end of this regrettable attitude in circles where novelty is being generated as we speak. This short narrative invites us to answer the following question: Are we, as members of a species that can discriminate common good and common evil, condemned to make similar mistakes over and over again? And if so, for how long?

The biotechnology developed in the last five decades was propounded under the naïve assumption that it would "improve on Nature", in what became the weary, shifting timeframe of … "the next decade". This technology was advanced under a gene-centric, deterministic assumption that has been used to promote and fund both basic and goal (profit?)-orientated biological research. For more than 50 years, this policy has prevailed over an organic, evolution-based interpretation of the evidence, resulting in distortion of priorities. Signs of this misguided choice are now trickling down into the daily life of advanced and less-advanced societies all over the world, as pesticide-resistant strains of crops and pathogens. Instead of emphasizing prevention, policies have favoured "cures". Such socially targeted mistakes take years to diagnose and even longer to correct ….

Another non-minor incentive in applying a mostly reductionist philosophy in public health policy has been the benefits that American taxpayers would conceivably accrue from indirect investment in goal-orientated biological research. Genetically modified organisms and patient-specific therapies have been the purported altruistic motivations for this enterprise. The result of massive financial input into this so-called evidence-based technological project should surely have been improvement in the living standards of all segments of society. Have we, actors and promoters, fulfilled our respective responsibilities in safeguarding the public interest? And if not, why not?

Behind the overblown claims were the names of truly outstanding scientists who made valuable discoveries in the biological sciences, especially in genetics and molecular biology. Few constraints were suggested by those in a position to judge whether these biological manipulations of our shared environment were risk-free for the segments of the society destined to be directly or indirectly affected by them. As members of society, we all bear the blame for not reacting individually or collectively to this blatant attack on our long-term personal and social well-being. In hindsight, however, what was truly disconcerting was the lack of attention to the perils that irresponsible manipulation of the environment could have for all of us. Not even microbiologists and geneticists, closest to the issues and interests involved, were able to mount a strategy that would prevent the promotion of this Faustian deal. Perhaps it will be social historians who eventually tell us what else was going on in American society in the 1980s, so that future generations may not repeat the mistakes that are still being made. The changes introduced into our shared environment cannot be rolled back: now we can only try to understand why this dangerous biological gamble was conceived and tolerated.

The lack of concern and critical attention by scientists and regulators was equivalent to admitting that there was nothing wrong with selecting for genetically engineered pest-resistant variants without fully evaluating their environmental impact. Only a few scientists and laypersons objected and were liquidated as "luddites". The overwhelming reaction by scientists and technocrats was effectively muted either by self-censure or complicity. Responsible scientists may have justified their silence by the impossibility of predicting the outcome of a scientific experiment, and one may admit that this technology represented a "novel" environmental experiment. Meanwhile, under the capitalistic social construct of most northern hemisphere countries, "novelty" is equally exploited by responsible and irresponsible entrepreneurship. Regardless of their motivations, most major participants either remained silent or actively approved this attitude. Was this overall reaction, or lack of reaction, just plain ignorance and incompetence, or both?

Regardless of who the enablers of this miscarriage of science were (and still are), a single word sums up the attitude of designers and executors of these misguided ideas. That word is HUBRIS. Webster's dictionary defines hubris as "transgression of divine or moral law through ambition or one of the passions, ultimately causing the transgressor's doom …". Others may add the word GREED to characterize the whole process. One was led to believe that greed was linked to bankers and stockbrokers of the 1980s, when the notion that "greed was good" surfaced in the American vocabulary of the middle and lower classes. As everyone now knows, HUBRIS and GREED have extended their influence to all societies. Regrettably, this fact of life will not spare those who for lack of education were unaware of the existential risks we are all now running, those whose knowledge entitled them to raise the red flags that Mariano Bizzarri is now raising but remained silent, and finally, the future generations who will suffer the consequences of our collective lack of scientific and social responsibility.

The facts and ideas that Bizzarri has chronicled and judged in this thoroughly researched book reflect not only an analytical mind but also the importance of a milieu in which collaborators generate and criticize facts and ideas. This is what he has fostered at the Department of Experimental Medicine and Pathology, University of Rome *La Sapienza*. Professor Bizzarri, a polymath aware of the potential and limitations of a comprehensive understanding of biological complexity, has written a scholarly analysis of a subject of great current and future interest. His clear narrative is a constructive effort that places in perspective and extends human quests embodied in the names of philosophers, poets, composers of music, scientists and those who, in their varied personalized ways, make our human spirit worth preserving in a wisely managed world.

Can anything positive be drawn from this nightmarish scenario? As I tried to convey at the beginning of this Foreword, everything is new in the living world that we all share. Despite the (perhaps weary) advice to young people that they should become educated to achieve the usual things that our societies perceive as conducive to ever-elusive "success in life" (love, family, wealth and health, in whatever order one prefers), from now on we should insist on becoming an educated global society in order to preserve, or perhaps just prolong, the survival of the human species.

Prof. Carlos Sonnenschein
Tufts University, USA

Io sono il prologo[1]

Genetic engineering combines genes that Nature separated [...]
the ideology of unlimited mixing is contrary to the western
tradition. Whether it involves the ambition of Prometheus [...] or
banishment from the earthly paradise, the warning is the same:
beware he who crosses the line [...] Never has the lesson of the
great legends been more apt. Never has the warning been more
urgent. But who still meditates on the wisdom of antiquity?

Jean-Marie Pelt[2]

You are a scientist, and you are supposed to respect a natural
order in all things.

Alfred Hitchcock[3]

In recent years, genetically modified organisms (GMOs) in agriculture have been
the subject of a controversy involving the media, politics, consumer organizations,
citizens and the scientific community. The debate has not always been worthy of the
high stakes involved and unfortunately has not always exploited what research had
in the meantime discovered. This has been true both for sustainers of GMOs, who
usually take an ideological stance, avoiding the questions raised, and for certain
sectors of so-called militant environmentalism whose approaches are based on dis-
parate philosophical and ethical premises but lack scientific rigour. The triumphant
announcements of the former have not stood the test, nor have the catastrophes predicted by the latter proven to have a rational or experimental basis. These extreme
camps should bear in mind that shouting loudly adds nothing to scientific truth,
a truth that by nature is historical and relative but in its field is fully valid and
legitimate. Proponents of biotechnology in agriculture often cite long reference
lists carefully purged of anything that could put transgenic operations in a bad
light. We therefore decided that it would be appropriate to base this book almost
exclusively on official documents (FAO, EU, US National Academy of Science)
and with rare exceptions on articles published in peer-reviewed scientific journals,
which should be an inalienable reference for any discussion. We think this should

---

[1] "I am the prologue" from: *I Pagliacci*, Leoncavallo.
[2] J.-M. Pelt, *Plantes et Aliments Transgeniques*, Fayard, Paris (1998).
[3] From *Torn Curtain* (1966) by Howard Terpning directed by Alfred Hitchcock.

be to everyone's advantage, especially those media that occasionally reproduce nonsense obtained from an unknown source.

GMOs are again topical, now that the EU's moratorium on cultivation of GM plants has ended. The regulations are about to be re-examined, together with applications for the authorization of hundreds of GM products. The debate is therefore not academic but real and articulated on different planes – scientific, cultural, political, economic – each protagonist in its own field.

The question is essentially as follows: 15 years after the introduction of GM seeds and food, only four GM plants have conquered significant market positions (maize, cotton, rape and soy). Most other new organisms were found to have severe disadvantages, or are still projects, demonstrating the intrinsic complexity of the enterprise. Increasing doubts about the real economic and agricultural advantages of GMO have emerged. Little is known about their potential for upsetting environmental biodiversity in an irreversible way. Worries about their impact on human health have not been dispelled but have increased with the evident incapacity of current international standards to ensure controls and enforcement. As yet there are no reliable assessment methods or models, nor are there likely to be in the near future. This makes it impossible to evaluate the nature (qualitative) and degree (quantitative) of risk, and thus to make predictions based on risk/benefit ratios. In such a situation it seems prudent to resort to the precautionary principle, as in similar past situations, such as the epidemic of bovine spongiform encephalitis (BSE) and hormone contamination of meat, putting the question on a collision course with World Trade Organization (WTO) and US food and agriculture policy. The outcome is difficult to forecast, but it is clearly important to be prepared, and to do so it is necessary to promote serious discussion inside and outside the institutions.

Compared with a few years ago, the impression is that the institutions that ought to be responsible for the fate of European food and agriculture policy have other things on their minds. It is to be hoped that this does not become an alibi for "distraction" that biotech companies or other governments ably exploit to prompt (or impose) EU policy. We should be aware that the climate is changing, not only that of public opinion. Many regions of Italy have declared themselves GMO-free and similar choices have been made by certain US counties. In America and Europe there is growing consumer opposition to GM foodstuffs. This opposition has so far been ignored, exorcized or simply derided as modern superstition. However, the transgenic industry will have to convince consumers if they want to conquer markets, unless they opt for antidemocratic solutions, presenting consumers with a *fait accompli* by declaring an alleged but undemonstrated "substantial equivalence" of GMO and non-GMO, and destabilizing the production of traditional foods by increasing frequent "accidental" contamination, imposing the transgenic variant on markets without labelling or any way of tracing and monitoring, in line with *post-market surveillance* models.

The stakes are high and call for a scientific approach that excludes ambiguity and alibis. An editorial in *Scientific American* noted, "Unfortunately, it is impossible to verify that GM crops perform as advertised. That is because agritech companies have given themselves veto power over the work of independent researchers."[4]

However, a vast scientific literature has raised specific questions that governments, companies and scientists *have to answer*. Indignant tirades and partisan speculations are unacceptable. Transgenic supporters' accusations of unscientific arguments should be *returned to the sender*, and the discussion resumed on an equal footing of mutual attention and respect, without prejudice or ideological preconceptions.

I hope publication of this book by WIT Press can fulfil this need.

Mariano Bizzarri
Anzio (Rome), November 2011

---

[4] 'Do seed companies control GM crop research?', Editorial, *Sci. Am.* (August 2009), available at: http://www.scientificamerican.com/article.cfm?id=do-seed-companies-control-qm-crop-research.

1

Once upon a time, there were strawberries

To be recognized as men means admitting the right to imitate the gods.

E. Morin

Tired of the infinitely great and infinitely small, scientists dedicated themselves to the infinitely average.

E. Flaiano

Tomatoes No Longer Go Bad

Tomatoes no longer rot. They remain red, shiny and perfect, as if they were made of plastic. Potatoes have become stronger than Rambo and are no longer attacked by insects, not to mention Bt maize which kills its own parasites. These are the wonders of biotech.[1]

Are these things really true? Apart from the productivity of the new crops obtained from GM seeds, are there repercussions for the environment and human health? Proponents of transgene technology seem unsure, as a recent report of the National Research Council of the United States of America repeatedly underlines the following needs:

... enhance agricultural productivity in ways that also foster more sustainable agricultural practices, enhance the preservation of biodiversity, and decrease the potential for health problems that could be associated with some types of pest-protected plants [...] Priority [...] to the development of improved methods for identifying potential allergens in pest-protected plants [...] For some novel pest-protectants developed for future commercialization, longterm toxicity testing may be warranted [...] Assess gene flow and its potential consequences [...] Develop transgenic or other

---

[1] F. Angeli, Così la genetica è finita al supermercato *Il Giornale* (19 July 2000; our translation).

techniques that decrease potential for the spread of transgenes
into wild populations.[2]

The scenario is therefore nebulous and contradictory. Contrary to what the mass media lead us to believe, the introduction of GM crops and foodstuffs does not seem to be without its risks and dangers. It would be surprising if it were!

Tumultuous and often unpredictable advances of science and technology have helped to radically change the world and the idea we have of nature. The material world has been systematically dissected to obtain as detailed as possible an analytical picture, while barriers of space, time and organization of living material have been broken and surpassed. What was unthinkable a few years ago is now possible, opening horizons that *seem* to promise a new Eden for mankind. At the same time, however, elements suggesting that the scientific enterprise is not without its dangers have been emerging. The concrete possibility of epic changes in the environment in which we live – due for example to the greenhouse effect, nuclear contamination and chemical pollution – and lasting, widespread, unacceptable social and economic inequalities have not diminished but increased, while new perils, such as genetic contamination by GMO which could irreversibly upset whole ecosystems, have made their appearance. Scientific enterprise is itself subject to processes and tensions which are not homogeneous or monolithic, which make it impossible to predict future developments. However, science, especially biology, is clearly on the threshold of a revolution that will discard one or more paradigms, or in other words *those theoretical models that act as filters for cognitive perception and by which phenomena are framed coherently.*[3]

This is one reason why the attitude of the media towards scientific information has changed over the years; once ignored, science now often has front-page honours. However, something is not entirely convincing: why do we only read triumphant announcements about epic discoveries? The innovations are presented as glamorous, experiments are all decisive and "cutting-edge", links in an uninterrupted chain of successes, all-time records and extraordinary events. Nothing is said about what was behind these results, the work, contradictions, limits, difficulties and struggles. The picture of science is considered aseptic, far from the world of humans and their choices, fears and passions. Science as portrayed by newspapers is a hymn to the myth of progress, devoid of real or substantial concerns, apart from the need to achieve new goals every day. Seldom is anybody asking where we are going or why. This type of science, that relies more on the authoritative and

---

[2]Committee on Genetically Modified Pest-Protected Plants, Board on Agriculture and Natural Resources, National Research Council of USA, *Genetically Modified Pest-Protected Plants: Science and Regulation*, National Academy Press, Washington, DC (2000), p. 7 *et seq.*

[3]R. C. Strohman, 'The coming Kuhnian revolution in biology', *Nat. Biotechnol.* 15, 194–200 (1997).

charismatic opinion of the latest guru than on experiment, recalls the tragicomic Giovambattista Marino (1569) with his idea that "to amaze is the aim of science". Intrinsically complex issues that are still debated in the scientific community, such as the introduction of GMO foods, do not lend themselves to facile popular science and cannot be summed up with the irresponsible and ridiculous slogans we find in the mass media.

The questions raised by biotechnology (not limited to food, but including, for example, human cloning and gene therapy) not only concern scientific enterprises but also the world view and *view of man* they express.[4] This problem falls in the still confused and uncertain field of bioethics and raises the theme of the very identity of humankind and how it may change in coming decades. Strawberries were recently modified by biotechnology, through insertion of a gene from an Arctic fish that imparts resistance to freezing;[5] hopefully, it does not impart fish flavour or transform strawberries into something else! Can we entertain the same hopes for humans of the third millennium? Will humans defend their specificity or will they have to undergo "evolution" manipulated according to the interests of biotech multinationals? This question is taken up later in this book.

For the moment, it suffices to say that on the basis of current knowledge and the data produced by the science itself, the rationale of biotechnology is based on reductionist assumptions that are now inadequate due to the complexity of the relationship between genotype (the DNA sequence) and phenotype (shape and size) and are therefore unacceptably limited, first and foremost regarding their ecological impact and the safety of GM products.[6] The evidence for these concerns rests in the numbers that measure the questionable success of biotech in agriculture and food production. Despite enormous investments (denied to more promising sectors), 20 years of research have produced very few GM plants that have found a place in the market, while no transgenic animal, stable and capable of transmitting the new gene to its offspring, has yet lived more than a few months.

---

[4]These themes are at the centre of a hot and often bitter debate around modern culture. There is much literature and here we only indicate some of the more significant contributions: L. M. Silver, *Remaking Eden*, Avon, New York (1997); P. Spallone, *Generation Games*, Women's Press, Philadelphia, PA (1992); J. Sapp, *Beyond the Gene*, Oxford University Press, New York (1988).

[5]R. Hightower, C. Baden, E. Penzes, P. Lund & P. Dunsmuir, 'Expression of antifreeze proteins in transgenic plants', *Plant Mol. Biol.* 17, 1013–1021 (1991).

[6]The University *La Sapienza* (Rome) and the Italian Ministry for Agriculture jointly promoted a day dedicated to the study of GMOs in 2002 (*Organismi Geneticamente Modificati: Il Tempo delle Scelte*, a cura di M. Bizzarri, A. Laganà, S. Vieri, G. Grafico, Rome, 2003), and an analysis of the problems related to GMOs was recently published by the Istituto Nazionale di Ricerca per gli Alimenti e la Nutrizione (INRAN) which coordinated a comprehensive research programme involving major Italian researchers (*Agro-biotecnologie nel contesto italiano*, a cura di G. Monastra, INRAN, Rome, 2006). The emerging picture clearly showed that concerns about the risks related to use of the new biotechnologies, and the entity of those risks, are well founded.

Hungry for Business

What, then, was the hurry to impose GMOs, if not a prompt return on capital invested? Or should we believe that concern for "world hunger", a hypocritical *cliché*, has melted the hearts of scientists and corporations, inducing them to set aside any form of precaution in favour of a disinherited Earth? Actually, there is no economic or other reason that can justify GMOs as a necessity. In most cases this necessity is only alleged. Margaret Mellon, director of the agricultural and biotechnological programme of the University of Washington, writes:

> Obviously, we take risks all the time. But why are we tak-ing these risks? If we didn't have an abundant food supply, if we didn't have something like 300,000 food products on our shelves already, then we would have an argument for taking this society-wide risk. But we've got plenty of food. In fact, we've got too much. And although we have many problems associ-ated with our food system, they are not going to be solved by biotechnology.[7]

A commonly raised objection has been that the problem regards the third world, not industrialized countries; however, in this case the problem would be political, not scientific, being caused by unequal distribution of resources. Poor countries have neither the means nor the know-how to use technologies that would make them even more dependent on western food and agricultural industries, necessitating heavy expenditure and complete transformation of their production models. To help these nations, it would be sufficient to stop plundering their abundant (or meagre) wealth. In actual fact, the new biotech multinationals are about as concerned about hunger in Africa and Asia as the average citizen is in the fate of the poke root (*Pytolacca decandra*). As aptly observed in an editorial in *The Lancet*:

> Crops genetically modified to have reduced susceptibility to pests are promoted as a solution to low food yields in developing countries. The motive of these promoters is profit, not altruism. Monsanto, one of the largest developers of genetically modified crops, has developed a grain that gives an improved crop and is sterile, so instead of keeping back some seeds for the next year's sowing, farmers must return to the supplier for more. In view of this unbridled commercial approach to genetic modification, it is perhaps not surprising that companies have paid little evi-dent attention to the potential hazards to health of genetically modified foods.[8]

The stakes are quite different and have to do with the availability of germplasm, the genetic resources of species and ecosystems. Whoever controls germplasm

---

[7]M. Mellon, 'Does the world need GM foods?', *Sci. Am.* 223, 62–65 (2001).
[8]'Health risks of genetically modified foods', Editorial, *The Lancet* 353, 1811 (1999).

will control production and society as a whole. The battle with science on this subject (e.g. the shameful, incredible question of "patents on life") has little or nothing to do with it. Unfortunately, many scientists have nevertheless joined the cause, becoming part, and zealous sustainers, of a model of development created by modern industry.

The reasons that seem about to make the "dominant paradigm" implode are therefore only partially intrinsic to the scientific system and can be ascribed to the inadequacy of the world model elaborated in the last 30 years. Science should of course be defended, but one should be aware that different models of scientific knowledge coexist and are proposed, side by side. Internal and external contradictions mingle and are amplified, making it necessary to revisit scientific enterprise in order to recover an integrated view of phenomena; not only to foster a "systems biology" perspective, going beyond the limits of a reductionist approach, but also to incorporate ethical criteria in evaluating the biotechnological and ecological impact of its discoveries. The development of science cannot be regarded as harming human destiny and cannot be justified furtively, *a priori*, on merely gnoseological, or worse, economic grounds, especially where irreversible consequences are possible.

Ethics is not the servant of science. Today, an ideological superstructure is invading everything: it is time for science to go back to be the warrantor of truth. The superpower of technology is ushering in an era of "technoscience" with the capacity to co-opt and enchant not only Dr. Strangelove but also the conscience of the common citizen. Therefore it is time to address some relevant ethical challenges linked to the biotechnology adventure, namely stressing what Pelt and Séralini suggest in their "ethical oath": "I must be involved in learning about scientific advancement, and I will think about my investigations and their consequences".[9]

---

[9]J.-M. Pelt & G.-E. Séralini, *Après Nous le Déluge*, Flammarion-Fayard, Paris (2008).

2

The apprentice sorcerer's laboratory

To create is divine, to copy is human.

Man Ray

The gene is not so important.

A. Lima de Faria

The gene does not build the character, but makes the difference.
G. Sermonti

The Pages of the Book of Life

The biological information enabling cells to synthesize proteins and regulate the complex biochemical machinery of life is deposited in genes, nucleic acid segments of different lengths that together constitute deoxyribonucleic acid (DNA), the main constituent of chromosomes, and ribonucleic acid (RNA). Chromosomes can be defined as aggregates of genes disposed in two complementary strands known as helixes because they twist around each other in a way that won its discoverers, Watson and Crick, the Nobel prize in 1953.[1] The DNA in chromosomes, which are paired and vary in number in different species, is tightly packed, like the pages of a book. For cells to read them, the strands have to be unfolded so the nucleic acids can act as templates for the coding (synthesis) of a particular protein. The essential aspects of the mechanism are simple: DNA, consisting of bricks known as nitrogenous bases or nucleotides, opens in response to a series of stimuli, producing a single strand of RNA complementary to that of DNA. The RNA migrates from the nucleus into the surrounding cytoplasm which is enclosed by the plasma membrane. In its turn, RNA acts as a template for the synthesis of proteins, which like nucleic acids, are made up of amino acid bricks. Three adjacent nucleotides form a triplet which is identified by the initials of the bases (adenine, guanine, cytosine, thymine and uracil) and each triplet encodes (i.e. recognizes) a certain amino acid: the sequence of triplets forms a code that enables a protein to be assembled in sequence, one amino acid at a time. The set of triplets necessary to synthesize

---

[1] J. D. Watson & F. H. C. Crick, 'Molecular structure of nucleic acids. A structure for deoxyribose nucleic acid', *Nature* 171, 737–738 (1953).

a polypeptide or an enzyme is a completely functional unit defined as a *gene*. The number of genes varies from 2000 in bacteria which are probably the simplest form of life, to about 30,000 in humans.[2] This amazing mechanism has been called the *code of life* because it fulfils the biological need of cells to replicate, ensuring transmission of biological information to the progeny, which is the basis of inheritance. When a cell duplicates, as in asexual replication (replication not involving genetic material from another individual), the genetic material duplicated by the mother cell is transmitted to the two daughter cells. In sexual replication, the cell or zygote (a new individual) obtained by fusion of two germinal cells inherits genomic material (genome is the set of genes) from both parents. Each parent contributes half of its chromosomes which together reconstitute the full number of chromosomes, characteristic for each species.

Genetic engineering is the set of techniques by which the genetic material can be manipulated, removing genes, cutting them up, inserting DNA segments (transgenes) from other individuals or other species with the aim of giving a cell a protein, or in other words a specific biological function, via insertion or removal of a gene.[3] These premises suggested the possibility of modifying or improving a product by giving it properties that it did not normally have, properties that were intended to be beneficial. The idea was to *transform* plants, bacteria and animals to induce their cells to produce enzymes, hormones and specific proteins, that would give the new (GM) organism different functions and properties, such as resistance to pesticides, herbicides, cold and heat, until the line between fantasy and reality became blurred. Thus explained, it seems simple, perhaps too simple, because beyond the simplifications of this brief explanation, the process is really very complex indeed.

The Central Dogma of Biology

The ordered disposition of sequences of bases in groups of three (triplets) builds genes which are disposed along the chromosomes, collectively forming the genome of the cell, which through the mediation of RNA, translates information in the cell's cytoplasm where it is converted into proteins (enzymes, receptors, structural proteins, etc.) that have specific functions and together determine the phenotype and behaviour of the cell. The information necessary to control this complex biochemical machine resides entirely in DNA; by self-reproduction of DNA it is transmitted to the progeny. DNA is the only determinant of inherited characters, and, extensively, it has been considered the only depositary of "biological information".[4] This premise is known as the "central dogma of biology" and is grossly misleading.

---

[2]The number of genes is considered as the number of DNA segments coding proteins.
[3]S. M. Kingsman & A. J. Kingsman, *Genetic Engineering*, Blackwell Scientific Publications, Oxford (1988).
[4]J. D. Watson, *A Passion for DNA: Genes, Genomes, and Society*, Cold Spring Harbor Laboratory Press, Cold Spring Harbor, NY (2000), pp. 223–225.

This theory – universally accepted until about 10 years ago – is now finally collapsing under the weight of scientific evidence, particularly data emerging from the Human Genome Project. Today we know that there are *too few genes* to explain the variety of hereditary characters and more generally the complexity of living organisms.

The aim of the Human Genome Project was to find all the genes in human cells by identifying the sequences formed by about three billion nucleotides. James Watson called this project the "definitive description of life", which would provide the information necessary to distinguish a carrot cell from that of a fly or a human. Even more significantly, Walter Gilbert, a major promoter of the project, hoped to be able to put the sequences on a compact disc and say, "Here is a human being, this is me!"[5] These hyperboles can be understood in the light of the assumptions and inevitable logical consequences of the theory formulated lucidly and coherently by F. Crick. DNA exerts control on hereditary characters through proteins that are structurally obliged to fulfil precise functions. The ability of DNA to govern synthesis of proteins is related to structural similarities between the two types of molecule: as underlined by Crick, the nucleotide sequence of the gene is the simple code for determining the amino acid sequence of a given protein.[6] In other words, there is perfect one-to-one correspondence between gene and coded protein, such that the set of genes reflects a similar number of proteins. Since DNA of all living cells is made up of the same nucleotides, the genetic code is universal. This means that a gene produces its "specific" protein wherever it is "activated", even in cell contexts very different from the original one, at least according to the original theory of Crick, which finishes with a statement aptly known as the "central dogma" of biology: once digital ("sequenced") information is translated into a protein, it cannot be modified. This means that information arising from the nucleotide sequence of DNA is translated unchanged into the corresponding amino acid sequence. This assertion is crucial to the classical theory of molecular biology, establishing a rigid one-to-one connection between DNA transcription (DNA is copied to a strand of messenger RNA (mRNA)) and its product (a protein) which is in turn responsible for the characters and specific functions a cell is considered to transmit to its progeny. The whole edifice rests on this foundation. Crick was lucidly aware of the need to postulate a one-way flow of biological information – from DNA to proteins via RNA – to give the theoretical model descriptive and predictive power. Any other hypothesis, such as considering the possibility of a network connection with interconnected circular two-way flows of genetic information would have mined the foundations of his theory: the discovery of even a single type of cell in which

---

[5] W. Gilbert, A vision of the Grail, in: *The Code of Codes: Scientific and Social Issues in the Human Genome Project*, D. J. Kevles & L. Hoof (Eds), Harvard University Press, Cambridge, MA (1992), p. 96.

[6] F. H. C. Crick, Protein synthesis, in. *Symposium of the Society for Experimental Biology XII*, Academic Press, New York (1958), p. 153.

genetic information passed from proteins to nucleic acids or from one polypeptide to another would have destroyed the conceptual basis of molecular biology.[7]

These unmentionable fears turned out to be well founded. In the course of the 1990s, increasingly convincing scientific evidence showed that the proposed model was too rigid, mechanistic, reductionist and incapable of explaining many observed phenomena.[8,9] It had to be replaced with a so-called *flexible* genome system.[10] In the last 20 years, many scientific discoveries have accumulated in favour of a dynamic, non-linear interpretation based on networks of functional interconnection between molecules able to convey biological information.[11] Discovery of the mechanism of action of prions, the demonstration of alternative splicing, new insights into factors determining the tertiary structure of proteins, all shook the edifice of classical molecular biology.[12]

Paradoxically, however, the *coup de grace* for the hegemonic ambitions of the central dogma of molecular biology came from the project that aimed to consolidate and confirm, once and for all, the *omnipotent gene* myth. Publication of the results of the two research teams involved in decoding the human genome (Human Genome Project[13]) confirmed their worst fears. Instead of the hundred thousand odd genes predicted by the theory on the basis of the number of known proteins, only 30,000 genes were identified.[14] This number is only two times the number of genes in the fruit fly and only 2% more than the genes of the mouse. On the basis of this "heritage" Barry Commoner pointed out that it would be difficult to distinguish a human from a non-human,[15] and indeed it would be easy to confuse Walter Gilbert with a mouse, since 99% of his genes matched those of the animal counterpart.[16]

This surprising result, which had indeed been anticipated by certain sectors of the scientific community, labelled as heretics and marginalized because they criticized the dominant theory of molecular biology, shook Crick's theoretical

[7]F. H. C. Crick, 'The central dogma of molecular biology', *Nature* 227, 561–563 (1970).

[8]H. V. Westerhoff & B. O. Palsson, 'The evolution of molecular biology into systems biology', *Nat. Biotechnol.* 22, 1249–1252 (2004).

[9]F. Mazzocchi, 'Complexity in biology', *EMBO Rep.* 9(1), 10–14 (2008).

[10]R. J. Greenspan, 'The flexible genome', *Nat. Rev. Genet.* 2, 383–387 (2001).

[11]R. Strohman, 'Manoeuvring in the complex path from genotype to phenotype', *Science* 296, 701–703 (2002).

[12]H. H. Q. Heng, 'The conflict between complex systems and reductionism', *JAMA* 300(1), 1580–1581 (2008).

[13]J. C. Venter, M. D. Adams & E. W. Myers, 'The sequence of the Human Genome', *Science* 291, 1304–1351 (2001).

[14]J. M. Claverie, 'What if there are only 30,000 human genes?', *Science* 291, 1304–1351 (2001).

[15]B. Commoner, 'The roles of deoxyribonucleic acid in inheritance', *Nature* 203, 486–491 (1964).

[16]B. Commoner, 'Unraveling the DNA myth', *Seedling* (July 2003), GRAIN, available at: http://www.grain.org/seedling/?id=240.

edifice,[17] and posed insoluble problems, though possible explanations are emerging with some difficulty.

First of all, today we know that coding DNA (about 2% of the whole genome) does not act in isolation but establishes relations of mutual collaboration with other apparently *mute* gene segments. A few years ago, light was shed on the crucial functions of what was hitherto arrogantly called "junk DNA" and which is now more hypocritically known as non-coding DNA.[18] "Junk DNA" is composed of short (SINE) and long (LINE) sequences. The latter are derived from the replication of mobile elements (transposons and retroposons) recognized by Barbara McClintock as long ago as 1952, but rejected by scientific orthodoxy. This discovery and its author were relegated for decades to the limbo of heresy.[19]

Classical theory teaches that mRNA translates the DNA sequence unchanged into a distinct amino acid sequence for each protein. However, many studies have shown that the information of DNA is split into many fragments that are recombined in different ways to produce two or more different proteins. The mechanism is based on interactions that a specialized group of proteins and certain small mRNA sequences, known as spliceosomes, establish with the nascent mRNA molecule.[20] This is cut and rebuilt, giving rise to one or more alternative combinations of mRNA which then govern transcription of the same number of different proteins.[21] Although the first transcription is essential and important (depending solely on DNA and its transcription in nascent mRNA), it is the spliceosomes system that determines the final message.[22] Alternative splicing means that a single gene in cells

---

[17]The press and much of the scientific community does not seem to realize what happened. However, significant public statements have been made by E. Lander, one of the directors of the Genome Project, and newspapers such as the *New York Times*, which reported that the affair taught a lesson in humility that will not be easy to forget.
[18]This also turned out to be false: it was recently discovered that much "junk DNA" is transcribed in proteins (see Fantom Consortium – Riken Consortium, 'The transcriptional landscape of the mammalian genome', *Science* 309, 1559–1563 (2005)).
[19]At the end of the fifties, McClintock questioned the stability of the genome on the basis of observations in maize. This won her the hostility and ostracism of most of the scientific community. Her work was finally appreciated in 1973, after Shapiro confirmed the existence of transposons also in bacteria (J. A. Shapiro, (Ed.), *Mobile Genetic Elements*, Academic Press, New York (1983); J. A. Shapiro, 'DNA insertion elements and the evolution of chromosome primary structure', *Trends Biochem. Sci.* 2, 622–627 (1877)).
[20]C. A. Collins & C. Guthrie, 'Allele specific genetic interactions between Prp8 and RNA active site residues suggest a function for Prp8 at the catalytic core of the spliceosome', *Genes Dev.* 13(15), 1970–1982 (1999).
[21]A linguistic example: consider the letters of the word TIME: these are transcribed by DNA and the spliceosome could assemble them in different ways to form the words ITEM or TIME, which have different meanings though formed from the same pieces (and quantity, i.e. bits) of information.
[22]C. A. Collins & C. Guthrie, *op. cit.*

of the human inner ear can give rise to 576 different proteins,[23] whereas only three genes code up to 2,258 different polypeptides[24] in the neurexine category (neuron connection proteins). In the fruit fly, there may be up to 38,012 variants of the same protein.[25] Rearranging the nucleotide sequence transmitted by a single gene in a different way, the spliceosome gives rise to a multitude of mRNA that generates new and different genetic information, not only with respect to that transmitted by the original DNA sequence but also with respect to the cell context: the same gene, in different cells, can be processed by different spliceosome systems, giving rise to alternative mRNA transcripts. The same genes may be transcribed beginning and ending at different points, and thus coding for more than one RNA per gene and giving rise to more than one protein.

These findings show that determination of a protein sequence does not reside exclusively with DNA, invalidating many of the aspirations underlying the Genome Project: they diminish the importance of the single gene in determining inherited characters and more generally in determining the structure, conformation and biological function of the single protein. In other words, the effects of the gene cannot be simplistically and linearly deduced from its nucleotide sequence, determination of which was the main objective of the Genome Project.

This came as no surprise to the scientific community. What is surprising is that the Human Genome Project could be planned without considering what was already known about alternative processing of genetic information. The Genome Project does not even mention alternative splicing, though this mechanism was demonstrated in 1978 in viral replication[26] and in 1981 in human cells.[27] At the end of 1989, when the project was developed, at least 200 scientific papers had already been published on the subject. One may well ask whether some other project was not behind the emphasis and obstinacy with which the project was proposed and sustained, a project the results of which, though important, could never have been what they were made out to be in the ridiculous, hysterical and misleading publicity campaign.[28]

---

[23]D. L. Black, 'Splicing in the inner ear: a familiar tune, but what are the instruments?', *Neuron* 20(2), 165–168 (1998).
[24]M. Missler, W. Zhang, A. Rohlmann, G. Kattenstroth, R. E. Hammer, K. Gottmann & T. C. Südhof, 'Neurexins couple Ca^{2+} channels to synaptic vesicle exocytosis', *Nature* 423, 939–948, (2003).
[25]D. Schmucker, J. C. Clemens, F. Shu, C. A. Worby, J. Xiao, M. Muda, J. E. Dixon & S. L. Zipursky, '*Drosophila* Dscam is an axon guidance receptor exhibiting extraordinary molecular diversity', *Cell* 101(6), 671–684 (2000).
[26]R. Nevins & J. E. Darnell Jr., 'Steps in the processing of Ad2 mRNA: Poly(A) + nuclear sequences are conserved and Poly(A) addition precedes splicing', *Cell* 15, 1477–1493 (1978).
[27]F. M. DeNoto, D. D. Moore & H. M. Goodman, 'Human growth hormone DNA sequence and mRNA structure: possible alternative splicing', *Nucleic Acids Res.* 3, 3719–3730 (1981).
[28]G. L. G. Miklos, 'The Human Cancer Genome Project – one more misstep in the war on cancer', *Nat. Biotechnol.* 23, 535–537 (2005).

As indicated above, alternative splicing is not the only discovery that eroded the credibility of the central dogma of biology. A fundamental requisite of the original model of Watson and Crick was based on self-replication of the double helix of DNA, essential to ensure the exact reproduction of genetic material transmitted to daughter cells. The problem is that the DNA molecule does not "reproduce" itself at all, but accumulates an unacceptable percentage of errors. Normally, in the context of a fertilized oocyte (zygote), the process that leads from that cell to an adult organism with billions and billions of differentiated cells, the three billion nucleotides that compose the DNA molecule undergo billions of replications, during which the original structure and the sequence are conserved with high fidelity. The error rate (insertion of one nucleotide sequence instead of another) is about 1 in 10 billion nucleotides. Placing a single strand of DNA and a mixture of nucleotides in a test tube, one obtains a new DNA in which the error rate is 1 in 100 nucleotides! However, if certain enzymes and then repair enzymes are added to the mixture, the error rate drops first to 1 in 10 million and then to 1 in 10 billion. Not only are the enzyme systems (proteins) of the cell essential for faithful replication of genetic material, but we have known since the 1960s that they influence the nascent nucleotide sequence just as much.[29] The final genetic information does not emerge solely from the data transmitted by the nucleotide sequence of the original DNA, but also from its cooperation with the cell protein system, the so-called epigenetic system or epigenome.[30]

Epigenetic signals can drastically influence the health and characteristics of an organism and within certain limits can be transmitted to future generations (epigenetic inheritance) without altering the underlying DNA structure.[31] The complex mechanism of epigenesis is still far from understood, but it has already revolutionized classical DNA theory and our understanding of major diseases such as cancer. Epigenesis was completely unsuspected when the central dogma of biology was formulated, and is probably the mechanism behind the non-linearity of relationships between genotype and phenotype and unexpected pleiotropic phenomena resulting from transgenesis.[32]

The protein finally produced by this close cooperation between DNA and the epigenetic system is not yet a protein with biological function. It has to acquire a specific spatial conformation, known as *tertiary structure*, that guarantees the possibility of establishing certain interactions with other cell components and expressing its functional tasks. In the classical model of molecular biology, it is assumed that

---

[29]S. Spiegelman, 'An in vitro analysis of a replicating molecule', *Am. Sci.* 55, 221–264 (1967).
[30]S. Beck & A. Olek (Eds), *The Epigenome: Molecular Hide and Seek*, Wiley, New York (2003).
[31]G. Felsenfeld & M. Groudine, 'Controlling the double helix', *Nature* 421, 448–453 (2003).
[32]A. P. Wolffe & M. A. Matzke, 'Epigenetics: regulation through repression', *Science* 286, 481 (1000).

the linear structure of the nascent protein folds by itself and spontaneously acquires its tertiary structure. No scientific evidence for this assumption has been given and indeed, it raises many questions. Spontaneous processes have special connotations in physical chemistry, being characterized by positive entropy and random distribution. However in nature, things go quite differently: the choice of conformation is anything but random, being specific and limited to certain well-defined geometric forms.[33] Moreover, while it is true that certain proteins in solution "spontaneously" acquire the appropriate conformation, many others remain in their native state and do not acquire tertiary structure unless they come into contact with other specialized proteins (chaperones) that trigger folding and enable the polypeptide to acquire the form that permits it to express biological functions. Striking confirmation of this won W. Prusiner the Nobel Prize in 1990.[34]

A prion and the corresponding "normal" protein have the same amino acid sequence but different spatial conformation, demonstrating that tertiary structure is largely independent of the succession of amino acids coded by DNA, and depends on other partly unknown factors, including specific cell context. Even more importantly, prions can influence the three-dimensional conformation of the corresponding normal proteins and therefore give rise to transmission of biological information from protein to protein, challenging another of the postulates of the classical theory of molecular biology. In the test tube and *in vivo* in the brain, prions "transform" their normal counterpart making it acquire prion conformation. The normal protein thus "modified" promotes further conversions like a bad apple in a bin of good ones. This chain reaction is what occurs in BSE or mad cow disease.

The persuasive power of molecular biology ultimately lies in the idea that the entire set of so-called DNA-based instructions forms a "program" able to drive and organize all biological functions and structures in space and time. The attribution of control to the DNA was strongly reinforced by Monod, Jacob and Lwoff, who interpreted their work as evidence for the existence of a genetic program, an analogy explicitly based on comparison with an electronic computer. Specific instructions at the level of DNA could then be seen to "program" or control the development and behaviour of the organism.[35]

These ideas married well with the gene-centred theories of evolution and the metaphor of "selfish" genes, which relegated the organism to the role of a disposable

---

[33]M. Denton & C. Marshall, 'Protein folds: laws of form revisited', *Nature* 22, 410–417 (2001).

[34]K. Kaneko, D. Peretz, K. M. Pan, T. C. Blochberger, H. Wille, R. Gabizon, O. H. Griffith, F. E. Cohen, M. A. Baldwin & S. B. Prusiner, 'Prion protein (PrP) synthetic peptides induce cellular PrP to acquire properties of the scrapie isoform', *Proc. Natl. Acad. Sci. U. S. A.* 92(24), 11160–11164 (1995).

[35]F. Jacob, D. Perrin, C. Sanchez, J. Monod & S. Edelstein, 'The operon: a group of genes with expression coordinated by an operator', *C. R. Acad. Sci. Paris* 250, 1727–1729 (1960).

transient carrier of its DNA.[36] This framework was doubtless seductive, but for it to be correct, three conditions need to be satisfied. The first is that the relevant program logic should actually be found in the DNA sequences. The second is that this should control the production of proteins. The third is that this should be a determinate process. Yet, none of these conditions have been fulfilled.[37] The C, G, A, T sequences of nucleotides in the genome do not themselves form a program as normally understood, with complete logic of a kind that could separately run a computer. Instead, the sequences form a large set of templates that the cell uses to make specific proteins, and a smaller bank of switches, the regulatory genes, forming about 10% of human genes. But they require much more than the DNA sequences themselves to operate, since those switches depend on input from the rest of the organism and from the environment. Organisms are interaction machines, not Turing machines. There is therefore no computer into which we could insert the DNA sequences to generate life, other than life itself. Far from just being a transient vehicle, the organism itself contains the key to interpreting its DNA, and thus to giving it meaning.[38]

The way to save the genetic program idea would be to abandon the identification of genes with specific sequences of DNA alone and return to the original idea of genes as the causes of particular phenotypes only by including other relevant processes in the organism. The problem with this approach is that the closer we get to characterizing the "program" for a particular phenotype, the more it looks like the functionality itself. Thus, the process of cardiac rhythm can be represented as such a "program" (indeed, modellers write computer programs to reproduce the process), but it is not a sequence of instructions separate from the functionality itself.[39] Therefore, there is no Genetic Program and there is no privileged level of causality in biological systems. DNA is a database for the transmission of successful organisms, rather than a "program" that "creates" them. Organisms are not simply manufactured according to a set of instructions. There is no easy way to separate instructions from the process of carrying them out, to distinguish plan from execution.[40] So far, genomes neither contain the future of the organism, in some pre-formative version of the van Leeuwenhoek's homunculi, nor are they to be regarded as blueprints or information theorists' code bearers.[41]

---

[36]R. Dawkins, *The Selfish Gene*, Oxford University Press, Oxford, UK (1976, 2006).

[37]D. Noble, 'Claude Bernard, the first systems biologist, and the future of physiology', *Exp. Physiol.* 93, 16–26 (2008).

[38]Y. Neuman, *Reviving the Living: Meaning Making in Living Systems*, Elsevier, Amsterdam, The Netherlands (2008).

[39]H. T. Siegelmann, *Neural networks and analog computation: beyond the Turing limit*, Birkhauser, Boston, MA (1998).

[40]E. Coen, *The Art of Genes*, Oxford University Press, Oxford, UK (1999).

[41]S. Roco, *Lifelines: Life Beyond the Gene*, Oxford University Press, New York (1997).

The lesson to be learnt from this is devastating for the central dogma of molecular biology. Namely, DNA has an important role but collaborates with many other molecules (messenger RNA, proteins and enzymes), as well as with cell structures, that not only take part in the repair and correction of translated sequences, but also insert new genetic information and help to transform the nascent strand of a protein into a correctly formed polypeptide with biological activity.[42,43] DNA by itself cannot be considered the source of genetic information translated into a protein, nor is it the source of the inherited character finally expressed by this protein. The correct replication and expression of DNA cannot be mechanistically ascribed to DNA alone, but it is performed by the concerted activity of various cell components. The information flows from DNA to RNA but also from RNA to DNA, or even between different RNAs.[44,45]

This reformulation of the conceptual basis of molecular biology has clearly had radical and unexpected effects, severely compromising the model on which modern biotechnologies are based. Genetic engineering presumes, without any convincing demonstration, that a foreign gene inserted in the chromosome of a host plant is capable of synthesizing exactly the same protein and nothing else.[46] We now know how great a misconception this has been. Through alternative splicing, one gene can produce thousands of protein variants, while almost unlimited, additional possibilities arise from RNA editing, translational regulation and post-translational modifications. The same proteins with the same amino acid sequences can, in different environments, be viewed as totally different molecules with different physical and chemical properties. Yet, we do not know how a foreign gene might interact with the host epigenome and spliceosome. In the new cell context, alternative splicing may give rise to multiple variants in the protein[47] or the latter may

---

[42]B. Commoner, 'Failure of the Watson–Crick theory as a chemical explanation of inheritance', *Nature* 220, 334–340 (1968).

[43]L. Moss, *What Genes Can't Do*, MIT Press, Cambridge, MA (2003).

[44]H. M. Temin & S. Mizutani, 'RNA-dependent DNA polymerase in virions of Rous sarcoma virus', *Nature* 226, 1211–1213 (1970).

[45]G. J. Hannon, 'RNA interference', *Nature* 418, 244–251 (2002).

[46]Statements by industry representatives indicate that biotech companies are founded on the classical assumption of molecular biology: omnipotence of the gene according to the central dogma. In 1999, during a US Senate hearing, Ralph W. Hardy, president of the *National Agricultural Biotechnology Council*, summarized the guiding theory of the biotech consortium in the following way: DNA (top management molecule) directs the formation of RNA molecules (intermediate management molecule) which in turn direct the production of proteins (worker molecules) (Report released by the *National Agricultural Statistics Service*, Agricultural Statistics Board, US Department of Agriculture, Acreage June 29, 2001). This management version of the central dogma would be puerile and ridiculous were it not the basis of a project that will wreak havoc on the planet's agriculture.

[47]M. T. Cheah, A. Wachter, N. Sudarsan & R. R. Breaker, 'Control of alternative RNA splicing and gene expression by eukaryotic riboswitches', *Nature* 447, 497–501 (2007).

acquire conformations different from the original one due to diversity of host chaperones[48] and hence unexpected and unpredictable biological functions that could be associated with completely new health and ecological risks. What determines which proteins are made and in what quantity is not the DNA alone. These instructions come from the cells and tissues themselves, in the form of varying levels of transcription factors, biophysical and epigenetic cues that are specific to the different types of cell.[49] Additionally, many genetic changes, either knockouts or mutations, appear not to have significant phenotypic effects; or rather they have effects that are subtle, often revealed only when the organism is under stress. For example, complete deletion of genes in yeast has no obvious phenotypic effect in 80% of cases. Yet, 97% have an effect on growth during stress. The reason is that changes at the level of the genome are frequently buffered, i.e. alternative processes kick in at lower levels (such as gene–protein networks) to ensure continued functionality at higher levels (such as cells, tissues and organs).[50]

Modulation of genetic heritage is strictly cell- and species-specific. The same gene in two cells from different tissues, even from the same organism, can give rise to different proteins[51] or may not be expressed at all. Well aware of the latest discoveries and that genetic exchange in nature (or during traditional crosses) only occurs within a given species, it is paradoxical that the biotech industry has promoted the erroneous view that transfer of a gene from one species to another, or even from one kingdom to another, is a specific and precise procedure.

The fidelity of reproduction of a particular protein in the context of a non-GM cell system depends essentially on compatibility coordinated between the gene and the protein transcription and translation machinery. The harmony between the different actors of this process, acquired in the course of a long evolutionary process, is threatened by intrusion of a gene segment that to be expressed must necessarily entrust itself to an unknown protein machinery. If a bacterial gene is inserted in a plant, nothing ensures that it will be correctly expressed. As a consequence, "in a transgenic plant, the harmonious interdependence between the alien gene and the protein system of the host is presumably altered in a way that is not specific,

---

[48]"Chaperones" are proteins that guide tertiary folding of nascent proteins. Proteins synthesized in "alien" contexts may have anomalous configurations, as in the case of prions, because they do not interact with the right chaperones (see R. J. Ellis & S. M. Hemmingsen, 'Molecular chaperones: proteins essential for the biogenesis of some macromolecular structures', *Trends Biochem. Sci.* 14(8), 339–342 (1989)).

[49]J. Maynard Smith & E. Szathmáry, *The Major Transitions in Evolution*, Oxford University Press, Oxford, UK (1995).

[50]M. E. Hillenmeyer, E. Fung, J. Wildenhain, S. E. Pierce, S. Hoon, W. Lee, M. Proctor, R. P. St. Onge, M. Tyers, D. Koller, R. B. Altman, R. W. Davis, C. Nislow & G. Giaever, 'The chemical genomic portrait of yeast: uncovering a phenotype for all genes', *Science* 320, 362–365 (2008).

[51]P. Comelli, J. Konig & W. Werr, 'Alternative splicing of two leading exons partitions promoter activity between the coding regions of the maize homebox gene Zmhox1a and Irap (transposon-associated protein)', *Plant Mol. Biol.* 41(5), 615–625 (1999).

imprecise and unpredictable",[52] irrespective of whether the transgenic operation was performed successfully.[53] Indeed, the same proteins with the same amino acid sequences can, in different environments, be viewed as totally different molecules with different physical and chemical properties.[54]

> The mastery of DNA acquired by researchers is by now evident. However, [...] an increasing number of new discoveries contradict the conventional notion that genes [...] encoding proteins constitute the sole foundation of inheritance and the entire blueprint of all living beings [...] Beyond the sequence of DNA in chromosomes there is another layer of information [...] namely epigenetic markers incorporated in a mixture of proteins and chemical substances that surround, sustain and attach themselves to DNA, operating through obscure codes and mechanisms [...] it will take years, perhaps decades, to build a detailed theory that explains how DNA, RNA and epigenetic mechanisms combine to form a self-regulating system. But there is no doubt that a *new theory is needed to replace the central dogma* on which molecular genetics and biotechnologies have been based since the fifties.[55]

This comment was published in *Scientific American*, an authoritative journal without environmentalist leanings, which gives voice to many proponents of GMOs. But are the proponents aware of these statements?

One Gene, One Enzyme

Genetic engineering is based on relatively recent knowledge, which everyone agrees is incomplete.

> We now know that things are much more complicated than we thought (and still think, according to the mass media and current school textbooks). This ignorance is dangerous [...] [because] it transmits an over-simplified view of life and gives the impression that we have the tools to change it at a whim, easily and without any risk of unexpected effects [...] [whereas] it is this very unpredictability that enables [living systems] to

[52] V. G. Pursel, R. E. Hammer, D. J. Bolt, R. D. Palmiter & R. L. Brinster, 'Integration, expression and germ-line transmission of growth-related genes in pigs', *J. Reprod. Fertil. Suppl.* 41, 77–87 (1990).

[53] V. G. Pursel, C. E. Rexroad Jr, D. J. Bolt, K. F. Miller, R. J. Wall, R. E. Hammer, C. A. Pinkert, R. D. Palmiter & R. L. Brinster, 'Progress on gene transfer in farm animals'. *Vet. Immunol. Immunopathol.* 17(1–4), 303–312 (1987).

[54] S. Rothman, *Lesson from the Living Cell: The Limits of Reductionism*, McGraw-Hill, New York (2002).

[55] W. W. Gibbs, 'The unseen genome: gems among the junk', *Sci. Am.* 289, 46–53 (2003).

> continue to live, adapting and changing in order to remain the
> same.[56]

The underlying assumption, obtained directly from microbiology, is that a single
gene expresses a well-defined function governed by a protein, an enzyme that
the same gene synthesizes from the complete instructions in its possession. This
conviction suggested that it was sufficient to modify one or more particular genes,
changing, removing or replacing them, in order to modify or regulate a function.
Biotechnology is based on this simplified equation of the world: one gene = one
enzyme. Now this postulate has to be revised. Cell level functions, and to an even
greater extent the functions of organs and whole organisms, are subject to many
regulating factors that depend on the sequential and coordinated activation of a
multiplicity gene sequences and on the role of other elements, not all of which are
necessarily genetic.

> DNA does not prepare anything, but provides the cell with
> information that is expressed selectively in relation to context.
> Morphogenesis is therefore a process that has various levels of
> uncertainty.[57]

This is illustrated by the fact that the tertiary structure of proteins (that enables the
proteins to act properly) is not determined by DNA but by interactions between
the polypeptide's microenvironment and its primary amino acid sequence.[58] The
living form – the so-called phenotype – thus arises from a complex interaction that
allows DNA and its external and internal environment to entertain a continuing
dialogue, constrained by an indefinite but limited number of possibilities.

According to Pollack, we will not discover the meaning of a gene by playing
with it in our labs or on our computers: we have to put it in a cell and let it show
us what it means. The first experiments showed that context is as critical for the
meaning of a gene as it is for the meaning of a word: a gene can have two completely
different meanings in two different cells or even in the same cell at different times.
To translate the complete meaning of even a single gene, we would need to see
the structure of the cells in which it is expressed, in various parts of the body
and at different times, as well as subsequent interactions of its proteins with other
molecules or cells.[59]

---

[56]M. Buiatti, Elementi di strutture e dinamica dei genomi, in: Agro-biotecnologie nel
Contesto Italiano, G. Monastra (Ed.), INRAN, Rome (2006), pp. 23–31.
[57]G. Corbellini, Le Grammatiche del Vivente. Storia della Biologia Molecolare, Laterza,
Bari (1997), p. 135.
[58]J. Rumbely, 'An aminoacid code for protein folding, Proc. Natl Aacd. Sci. U. S. A. 98(1),
105–108 (2001).
[59]R. Pollack, Signs of Life. The Language and Meanings of DNA, Houghton Mifflin,
New York (1994).

Today it is less certain that such superb harmony is the product of chance and perpetuates through the necessity of survival than it was when Monod sent his pessimistic *Chance and Necessity* to press. The words of Kepler in distant 1606 are food for thought:

> Yesterday, tired of writing, I was called to the table and was served the salad I had requested. I said to myself, "Is it possible that plates, lettuce leaves, grains of salt, drops of oil and vinegar and slices of egg, flying in the air for all eternity, combine by chance, forming a salad?"
>
> "Yes," replied my love, "but not as good as mine!"[60]

How could it be otherwise! After all, Hamlet warned Horace that the world was much more complex than current knowledge could admit: "There are more things in heaven and earth, than are dreamt of in your philosophy."

Moreover, genes of any species work under complex systems, ruled out by non-linear dynamics, in a coherent network of interactions with their environment. This implies that the "morphogenetic field" of the "complex system" (cell and surrounding stroma) is involved in modulating the expression and "functioning" of the genome.[61] No gene works in isolation and no gene determines *per se* the characters of the phenotype in a direct linear manner. The proof lies in the fact that the *differences* in terms of genes content between a human and a mouse are less than 2%. An important observation was made by F. Jacob in distant 1977:

> Species differ more in the chemical functions activated or silent in them than in DNA constitution. What distinguishes a butterfly from a lion, a hen from a fly or a worm from a whale is much less a difference in chemical constituents than in distribution and organization of these constituents [...] Specialization and diversification took place using the same information *in different ways*.[62]

The simplistic equation of classical molecular biology ("one gene, one protein, one function"), a paradigmatic example of reductionism and biological mechanism, has finally lost its allure and, above all, its validity.[63] The expression of a gene varies from individual to individual, and from tissue to tissue, depending

[60] J. Kepler, *De Stella Nova* (1606), quoted in G. Sermonti, La costola d'Adamo, *Riv. Biol.* 77(2), 12 (1984).

[61] S. Dinicola, F. D'Anselmi, A. Pasqualato, S. Proietti, E. Lisi, A. Cucina & M. Bizzarri, 'A systems biology approach to cancer: fractals, attractors, and nonlinear dynamics', *OMICS* 15(3), 93–104 (2011).

[62] F. Jacob (1977), quoted in G. Sermonti, L'organismo strutturale, *Biol. Forum* 91(1), 8 (1998).

[63] D. Noble, 'Genes and causation', *Phil. Trans. R Soc. A* 366, 3001–3015 (2008).

on the genetic background and the biochemical and physicochemical characteristics of the microenvironment. Dynamic interactions between genome and cell
environment make DNA a flexible structure that can be modulated within wide
limits. Stabilization of gene function arises from this continuous dialogue with the
biophysical constraints ("forces"), shaping the field in which the biological system
belongs.[64,65] Disruption of this field might likely induce phenotypic as well as pathological changes.[66,67] Indeed, by considering the origination and transformation of
developmental mechanisms as an evolutionary problem in its own right, we observe
that non-genetic mechanisms, rather than genetic changes, are the major sources
of morphological novelty in evolution:[68] (1) interactions of cell metabolism with
the physicochemical environment within and external to the organism, (2) interactions of tissue masses with the physical environment on the basis of physical laws;
(3) interactions among tissues themselves, according to an evolving set of rules.
The forms and characters assumed by metazoan organisms originated in large part
by the action of such processes.[69] Thus, insertion of even a single foreign sequence
may cause the organism to change in a non-linear and unpredictable way. In other
words, it is not possible to predict the consequences of a gene transplant *a priori*.
The same concerns hold for any manipulation of this type and hence also for the
recently popular idea of gene therapy.

We still know very little about DNA, indeed, no function has yet been recognized
for more than 97% of genes. This "mute" DNA that bioengineers arrogantly refer
to as junk must nevertheless have some important role. Nature is parsimonious and
does not seem to do anything by chance. According to Einstein, she hides her secret
out of substantial superiority, not out of astuteness. To presume that there is no
harm in introducing new units into this "mute" mass because we do not know the
significance of most of the chromosome sequences is like saying that no harm can
come from not knowing what we are doing.

---

[64]M. Levin, 'Bioelectomagnetics in morphogenesis', *Bioelectomagnetics* 24, 295–315 (2003).

[65]S. A. Newman, 'Developmental mechanisms: putting genes in their place', *J. Biosci.* 27(2), 97–104 (2002).

[66]M. Bizzarri & A. Giuliani, Representing cancer cell trajectories in a phase-space dia-
gram: switching cellular states by biological phase transition, in: *Applied Statistics
for Network Biology: Methods in Systems Biology*, M. Dehmer, F. Emmert-Streib,
A. Graber & A. Salvador (Eds), Wiley, Weinheim (2011).

[67]A. M. Soto & C. Sonnenschein, 'The somatic mutation theory of cancer: growing
problems with the paradigm?', *Bioessays* 26, 1097–1107 (2004).

[68]K. Saetzler K, C. Sonnenschein & A. M. Soto, 'Systems biology beyond networks: generating order from disorder through self-organization', *Sem. Cancer Biol.* 21(3), 165–174 (2011).

[69]S. A. Newman, G. Forgas & G. B. Muller, 'Before programs: the physical origination of
multicellular forms', *Int. J. Dev. Biol.* 50, 209–299 (2000).

Insertion of the Transgene

Insertion of new gene units involves risks that cannot be ignored. First of all, it causes momentary rupture of the double helix of DNA resulting in a new spatial arrangement of nucleic acid. This is a mutation *per se*, the effects of which we do not know and cannot predict. Secondly, the process of insertion of the new gene is often imprecise and is in any case random: the inserted segment may end up in the middle of a pre-existing sequence, altering its function and interfering with the activity of adjacent genes by contact.[70] The risk is considered low by spokesmen of the industries involved, but there is no scientific evidence to back up this claim. Indeed, it is certain that genes introduced artificially are subject to a higher frequency of spontaneous mutation, which is indirect evidence of the instability of such "crosses".[71]

Insertion of foreign genes is such an unnatural process that the cell systematically opposes it, trying to block incorporation of the new DNA segment or prevent its transcription by a biochemical operation that leads to methylation and hence inactivation of the "infecting" gene.[72,73] Insertion of the gene, especially if conveyed into the cell by a virus, may in turn alter the methylation process. This is serious interference in cell metabolism, inhibiting certain genes, selectively amplifying others and modifying complexes that may have a carcinogenic role.[74]

"Transplant" of the new gene disturbs the harmonious chain of processes that ensure not only homeostasis of the system but also its coherence, meaning a sort of "melody" that binds all the elements defining cell individuality together in coherent behaviour. Inserting a new gene has been compared to the surreptitious and arbitrary addition of a wrong note in a melody.[75] This metaphor explains a whole series of expected events, all of which have been observed.

First of all it explains why the success rate of GMOs is so low; the ratio of the number of attempts at inserting a gene and the number of resulting GMOs capable of transferring the new characteristics to their progeny is surprisingly

---

[70] G. M. Whal, 'Effect of chromosomal position on amplification of transfected genes in animal cells', *Nature* 307, 516 (1984).

[71] P. R. Beetham, P. B. Kipp, X. L. Sawycky, C. J. Arntzen & G. D. May, 'A tool for functional plant genomics: chimeric RNA/DNA oligonucleotides cause *in vivo* gene-specific mutations', *Proc. Natl. Acad. Sci. U. S. A.* 96, 8774 (1999).

[72] H. Heller, 'Chromosomal insertion of foreign DNA is associated with enhanced methylation of cellular DNA segments', *Proc. Natl. Acad. Sci. U. S. A.* 92, 5515 (1995).

[73] S. P. Kumpatla, W. Teng, W. G. Buchholz & T. C. Hall, 'Epigenetic transcriptional silencing and 5-azacytidine-mediated reactivation of a complex transgene in rice', *Plant Physiol.* 115, 361–373 (1997).

[74] W. Doerfler, 'Patterns of DNA methylation – evolutionary vestiges of foreign DNA inactivation as a host defense mechanism', *Biol. Chem. Hoppe Seyler* 372, 557–562 (1991).

[75] J. G. Perez, *Planète Transgénique*, L'Espace Bleu, Paris (1997).

low: for plants it is about 1:1,000; for animals it varies widely. On average, only 10–25% of modified embryos proceed to birth, and the probability that the modified gene is present in the progeny is even lower (0–20%). Better results are obtained with small animals (5% success rate for transgenic mice), and poorer results with large ones (only 1 in 100 or 1,000 embryos results in a GM sheep, goat or cow).[76] The balance, drawn up backstage of the triumphant publicity, is devastating:

> In nearly 20 years only two well-defined characters have been successfully modified in food plants [...] through the insertion of few genes; and, until now, no transgenic animal has entered the market place. Obviously, this did not means that a lot of both plants and animals have been modified during laboratory experiments involving several hundred of genes; this signify that in the latter case no one of the modified vegetables has been considered adequately productive or safe to enter the market.[47]

Not only is it difficult to insert a foreign gene and make it express the desired character (the protein or enzyme), since this also depends largely on the biochemical and structural context in which the new gene has to operate, but it is also highly improbable that the new character be conserved in the progeny without giving rise to a series of processes that culminate in instability of the host genome. Many biologists sustain that gene transplants favour destructuring of DNA and increase the probability of spontaneous mutations, making transgenic cell lines undergo further unpredictable and uncontrollable modifications.[49]

Several mechanisms capable of explaining this high rate of variability have already been identified. Firstly, a similar situation is found in cell cultures: a cell is explanted from its organism, its environment of physiological regulation, where many signals and factors work together to stabilize gene expression and modulation of functions, and is at high risk for genetic variations known as *somatoclonal variations*.[77] These are the reason why *Unilever*, a well-known Dutch firm, had to back out of a project to "regenerate" [*sic*] the plantations of palm oil in Malaysia.[78] The plants in which the gene of "long life" was inoculated turned out to be sterile when transplanted or died: another unexpected effect!

Secondly, it is impossible to determine where the new gene will be inserted in the double helix of DNA (position effect):[79] the transfected segment may end up in "mute" regions of the filament where it is not expressed; it may hook onto particularly fragile regions, causing mutations or DNA breakage, or even find itself near genes whose function it may inhibit, alter or modify; finally it may induce

---

[76]L.-M. Houdebine, *Le Génie Génétique. De l'animal à l'homme?*, Flammarion, Paris (1996).
[77]E. C. Cooking, Plant cell and tissue culture, in: J. L. Marx (Ed.), *A Revolution in Biotechnology*, Cambridge University Press, New York (1989), p. 119 *et seq*.
[78]OGM, Legambiente report, Rome (2000), p. 56.
[79]B. Lewin, *Genes VIII*, Longman Edition, Harlow, UK (2004), p. 512.

spatial remodelling of the helix with major repercussions on expression and on the function of the introduced segment. There are many possible situations: they are not understood and carry serious consequences. Despite the reassurances of the biotech companies, the products are anything but promising.[80]

As far as plants are concerned, these uncertainties are amplified by genome "fluidity",[81] largely related to the existence of transposomes. Transposomes were first identified in maize and are a reason for the genetic variability of this plant. The mobility of transposomes

> is accentuated when the plant is stressed, as when it is pro-
> cessed *in vitro*. It is therefore not surprising that when plant cells
> are maintained in active proliferation and cultured on synthetic
> media [...] they show enormous genetic variability.[82]

There is also an extraordinary increase in the frequency of gene mutations,[83] activation of transposomes[84] and methylation patterns.[85] It is surprising

> that for a long time, all this was considered unimportant for the
> production of GMOs having desired qualities and homogene-
> ity, at least from the point of view of genome structure. This
> denial of problems arising from the fluidity of plant genomes,
> which meant that the reactions of the plant in the presence
> of genes from other species were intrinsically unpredictable,
> delayed research into this crucial question [...] until the last ten
> years, and those regarding modification of transgene structure
> until the last five years.[86]

Finally, the modified gene can undergo a series of biochemical transformations (methylation, fragmentation) and rearrangements,[87,88] the consequences of which

[80] W. Doerfler, G. Orend, R. Schubbert, G. Fechteler, H. Heller, P. Wilgenbus & P. Schroer, 'On the insertion of foreign DNA into mammalian genomes: mechanism and consequences', *Gene* 157, 241 (1999).
[81] S. Brunner, K. Fengler, M. Morgante, S. Tingey & A. Rafalski, 'Evolution of DNA sequence nonhomologies among maize inbreds', *Plant Cell* 17, 343–360 (2005).
[82] M. Buiatti, *L'interazione con il genoma ospite*, in: *Agro-biotecnologie nel Contesto Italiano*, G. Monastra (Ed.), INRAN, Rome (2006), pp. 33–46.
[83] D. A. Evans & W. R. Sharp, 'Single gene mutation in tomato plants regenerated from tissue culture', *Science* 221, 949–951 (1983).
[84] V. R. Peschke & R. L. Phillips, 'Activation of the maize transposable element suppressor-mutator (Spm) in tissue culture', *Theor. Appl. Genet.* 81, 90–97 (1997).
[85] P. Bogani, A. Simoni, P. Liò, A. Germinario & M. Buiatti, 'Molecular variation in plant cell populations evolving *in vitro* in different physiological contexts', *Genome* 11, 1–10 (2001).
[86] M. Buiatti, *op. cit.*
[87] P. Windels, I. Taverniers, A. Depicker, E. Van Bockstaele & M. De Loose, 'Characterisation of the Roundup Ready soybean insert', *Eur. Food Res. Technol.* 213, 107–112 (2001).
[88] S. K. Svitashev, W. P. Pawlowski, I. Makarevitch, D. W. Planck & D. A. Somers, 'Complex transgene locus structures implicate multiple mechanisms for plant transgene rearrangement', *Plant J.* 32, 433–445 (2002).

are completely unpredictable[89] and include blockade of transgene expression,[90] or inactivation by operator and regulator genes, leading to silencing or functional inactivation. Indeed this is not a new discovery, but for some reason molecular biologists only seem to have "rediscovered" it recently.[91]

Obviously, all this is not only "theory" or considerations arising from some obscure experiment in the laboratory. Unfortunately it is solid reality. Just ask the producers of modified seeds of tobacco – they are well aware that 64–92% of first-generation transgenic plants later become unstable and have to be destroyed. This frequency, by the way, is anything but unexpected, since genomic instability ranges from 50% to 90% in plants such as *Arabidopsis thaliana* used as model for the study of plant DNA.[92]

Marcello Buiatti made an instructive and probably not exhaustive list of unexpected "alterations" related to insertion of transgenes, since "there have been various cases of plants already on the market that on closer analysis showed different modifications of inserted genes, forcing the owners to patent the new version or withdraw it from the GMO market".[93]

---

[89]A. Kohli, M. Leech, P. Vain, D. A. Laurie & P. Christou, 'Transgene organization in rice engineered through direct DNA transfer support a two-phase integration mechanism mediated by the establishment of integration hot spots', *Proc. Natl Acad. Sci. U. S. A.* 95(12), 7203–7208 (1998).

[90]S. Kumar & M. Fladung, 'Transgene repeats in aspen: molecular characterisation suggests simultaneous integration of independent T-DNAs into receptive hotspots in the host genome', *Mol. Gen. Genet.* 264, 20–28 (2000).

[91]B. Lewin, *op. cit.*, p. 530.

[92]Ho Mae-Wan, *Genetic Engineering Dreams or Nightmares. The Brave New World of Bad Science and Big Business*, Research Foundation for Science, Technology and Ecology at Third World Network (1997).

[93]M. Buiatti, *op. cit.* Buiatti lists an impressive series of "unexpected effects" including: "Maize, event MON863: besides a single copy of the transgene, it contains a copy of the marker NPTII and a fragment of the gene *ble*; Maize event MON810: in this case the variety was patented at least two more times after discovery of unexpected modifications [...]; maize event Nt176: three unexpected rearranged fragments were found; the first was of 118 base pairs and was homologous with the promotor 35S; the second contained a fragment of the same promotor and an unknown sequence of 215 base pairs; the third contained promotor P35S and the gene *bla* [...] these plants were withdrawn from the market; soy event 40-3-2 (Round-up-Ready): this variety has two extra fragments of the gene CP4EPSP for resistance to the herbicide [...] in 2000 Monsanto provided some extra information, admitting the presence of 254 bases of the above gene adjacent the known sequence and another 72 separate bases; independent research revealed a further fragment of 540 bases of unknown origin (P. Windels, I. Taverniers, A. Depicker, E. Van Bockstaele & M. De Loose, *op. cit.*). If this were not enough, it has been observed (A. Rang, B. Linke & B. Jansen, 'Detection of RNA variants transcribed from the transgene in Roundup Ready soybean', *Eur. Food Res. Technol.* 220, 438–443 (2005)) that at least 150 bases of the fragment [...] are transcribed together with the sequence of the resistance gene, giving rise (and here lies the most important novelty) to no less than four different RNAs that will give rise to completely unexpected and unknown proteins."

The mechanisms of modification and epigenetic modulation determine indirect, pleiotropic effects of transgenesis, not all necessarily with toxicological implications, with which molecular biology is increasingly concerned.[94] The emergence of these properties is substantially due to the fact that genes work in a complex network capable of amplifying or annulling even major perturbations linked to modification of one or more gene segments. This means that relations between genotype and phenotype – together with characters and functions expressed by the cell – are anything but linear and cannot be forced into the deterministic models typical of reductionist approaches.[95] This means that the substitution or deletion of a single gene may have no effect, or on the other hand have a cascade of unpredictable consequences.

The situation was effectively summed up in a letter by David Schubbert, professor at the Salk Institute of the University of California at La Jolla, published in *Nature Biotechnology*. The creation of a GMO raises three orders of concern, which have been given very little, if any, attention. First, the introduction of a gene into different cells of the same organism may give rise to different proteins. Furthermore, modifications mediated by the spliceosome and post-translational insertion of gycoside, phosphatide or lipid residues may alter the biological activity of the protein expressed by the transgene:

> With our current state of knowledge [...] there is no way of predicting either the modifications or their biological effects. Therefore, a toxin that is harmless to humans when made in bacteria could be modified by plant cells in many ways some of which might be harmful. [...] Introduction of one gene usually alters the gene expression pattern of the whole cell and typically each cell type of the organism will respond differently [...] Recent studies in transgenic plants showed that the overexpression of a gene involved in pectin synthesis had no effect on tobacco, but caused major structural changes and premature leaf shedding in apple trees [...] The introduction of genes [...] could lead to the synthesis of unexpected or even totally novel products through an interaction with endogenous pathways. Some of these products could be toxic [...] Prompt toxicity might be rapidly detected once the product entered the market place if it caused a unique disease, and if the food were labeled for traceability [...] However, cancer or other common diseases with delayed onset would take decades to detect, and might never be traced to their cause [...] The problem is, of course, that unless we know exactly what to look for, we are likely to miss the relevant changes.[96]

---

[94] P. Bettini, S. Michelotti, D. Bindi, R. Giannini, M. Capuana & M. Buiatti, 'Pleiotropic effect of the insertion of the *Agrobacterium rhizogenes rolD* gene in tomato (*Lycopersicon esculentum* Mill.)', *Theor. Appl. Genet.* 107, 831–836 (2003).

[95] U. Kuhnelin, R. Parsanejad, D. Zadworny & S. E. Aggrey, 'The dynamics of the genotype–phenotype association', *Poultry Sci.* 82, 876–881 (2003).

[96] D. Schubbert, 'A different perspective on GM food', *Nature Biotechnol.* 20, 969 (2002).

Fibonacci's Series

Too often molecular biologists sustain that the architecture of living material obeys the engineering logic developed by humans in the last two centuries. When phenomena cannot be framed in this simplistic and ingenuous mind-form, reckless solutions are excogitated. An example is "junk DNA", considered thus because its significance is not understood. The French mathematician Jean-Claude Perez, a collaborator of Luc Montaigner (who discovered HIV), set out to unravel the mystery and discovered a "secret" disposition of DNA: the sequence of genes follows the order of the Fibonacci series.[97] According to Perez:

> The architecture of DNA is a perfect clock in which even non-coding regions ("mute" regions as distinct from genes) take part, contrary to current opinion. When a foreign gene is inserted in DNA, this clock is broken: the gene enters somewhere in the genome, but we do not yet know how, where or why [...] we may trigger a cycle of mutations unrelated to each other and the mutation rate may increase. The genome may lose its stability and new viruses cannot be excluded. Modifying DNA is not like playing with *meccano*.[98]

We do not know the deep meaning of gene sequences in the configured space of the chromosome, nor do we know the role, not only and not so much in relation to expression of a function in an individual but in relation to the overall equilibrium of the germplasm of a species. There must be a reason, even if we do not know it, for the existence of this "extra" DNA and of genes we consider "harmful", since all together they make up the essential polymorphism of all living races.

According to Renato Dulbecco, polymorphism is a condition of evolution. The human species has a great number of genotypes, i.e. genetic variants, with evident differences.[99] It is only a short step from here to realizing that manipulation of plant and animal genomes could impoverish this variability, favouring genetic uniformity that would prejudice humanity's capacity to survive. However, bioengineers object that their operations are aimed at modifying segments that determine clear biological disadvantages, limiting the productivity of crops or responsible for diseases. They cite the example of thalassaemia (microcythemia), an inherited Mediterranean anaemia that has different forms of expression in relation to the type of genetic alteration which may involve one or more genes of one or more chromosomes, all however involved in the synthesis of haemoglobin, the protein that transports oxygen in red blood cells. Carriers of this genetic anomaly are undoubtedly unfortunate as individuals, developing different pathologies, one of which

---

[97] The rule of the Fibonacci series is that each number is the sum of the previous two, starting with one: 1-1-2-3-5-8-13-21-34-55-89, etc.
[98] J. C. Perez, *Planète Transgénique*, L'Espace bleu, Paris (1997), p. 67.
[99] R. Dulbecco, *Ingegneri della vita*, Sperling & Kupfer, Milan (1988), p. 54.

(Cooley's disease) is fatal. It would be a great advance to intervene on stem cells in the bone marrow to remove the "bad" gene and replace it with a "good" one. No problem – if such an operation were possible. However, it is worth asking why the group of alterations responsible for thalassaemia was conserved in the course of evolution, when evolution is known to get rid of useless or harmful genotypes. In a biological context, could these apparently harmful genes (and they are undoubtedly harmful for individual persons) have some benefit for the species? The name of the pathology should tell us something: thalassaemia or Mediterranean anaemia indicates that the genotype is prevalent among people living near water (*thalassa*) where thalassaemia exists alongside malaria. Today we know that the curves of incidence of these two diseases have inversely proportional profiles: when one increases, the other decreases and vice versa. By virtue of their smaller red blood cells (whence the name microcythemia), carriers of thalassacmia cannot host the plasmodium of malaria and do not contract malaria: *historically, the existence of a pool of thalassaemics prevented various forms of malaria from decimating populations living close to the sea.* What would have happened if we had removed the thalassaemia genes from the population? In other words, *who* decides whether a gene is harmful, and on the basis of what fragmentary knowledge?

Whether we like it or not, things are the way they are because they could not be otherwise. In the words of the biologist Giuseppe Sermonti:

> Among the infinite number of possible worlds, this is the only one that we can bear witness to. But does this not mean that it is the only world effectively possible? This may put forward again the debate around finalism or something even stranger: a world that goes backwards, whose explanation is not in the initial conditions but in the future. Science should therefore not expect humans to entrust the world to her, as if she had the logic to compose and manage it [...] The humility of Science brings the lost primate back to earth, allowing him/her to exist without asking permission. Stones, plants, animals and humans need not feel themselves to be the unlikely, arbitrary application of some provisional theory or the product of meaningless coincidences.[100]

---

[100] G. Sermonti, 'Il diapason dell'Universo', *Riv. Biol./Biol. Forum* 75(4), 6 (1982).

3

The ultragene invasion

The market must not be influenced by consumer choices
R. Shapiro, President of Monsanto

We cannot make blind bets on Nature: to save a tree or obtain
better flavoured fruits we risk upsetting an equilibrium that has
endured for millions of years
R. Dulbecco

The Devil Makes the Saucepans but Not the Lids (Italian Proverb)

The arguments used by companies engaged in the production of GM plant varieties
to justify their activity are based on three main claims: (1) transgenic crops have
selective and reproductive advantages, such as resistance to herbicides and parasites, compared to their natural counterparts; (2) the products are "substantially
equivalent" to natural foods, or even superior because they contain substances
and proteins of undeniable utility; (3) genetically altered crops give better yields,
enable fewer pesticides to be used and are a winning strategy for hunger in developing countries. The most optimistic add that it will soon be possible to invent plants
that produce drugs, vitamins and other substances, and that they may soon help to
defeat cancer.[1]

Such brave claims are inappropriate and not sustained by the facts, which to the
contrary indicate a high probability of danger for the environment and humans.
Transgenic crops have not proven to provide any substantial increase in yield,
which continues to vary in relation to many parameters, or to have a significantly
lower environmental impact,[2] as we shall see.

Producing More

A two-year study by the US government suggests that genetically
modified crops produce no better yields and require the use of
no less pesticide than non-GM crops.[3]

---

[1]S. G. Uzogara, 'The impact of genetic modification of human foods in the 21st century:
a review', *Biotechnol. Adv.* 18, 179–206 (2000).
[2]B. Johnson & A. Hope, 'GM crops and equivocal environmental benefits', *Nat.
Biotechnol.* 18, 242 (2000).
[3]*The Guardian*, 9ᵗʰ July 1000.

This is not an encouraging start, and unfortunately it has been confirmed by many studies. One of the first studies, conducted under suboptimal growing conditions in order to show even small significant differences, compared the yield of GM and non-GM soybeans. It found that the GM crop had yields that were 4–8% lower,[4,5] probably due to "unexpected effects" or "nodulation" of roots and reduced nitrogen-fixing capacity.[6] While significant differences in the yield of Bt-maize and Bt-cotton with respect to their non-GM counterparts were recently confirmed,[7] a significant increase in herbicide application (mean 5%[8]) was recorded for GM soybeans. American farmers who grew GM soybeans doubled the amount of herbicide employed each year.[9] Moreover, agricultural income does not seem to have benefitted from the introduction of the new varieties, since except for soybeans, the cost of transgenic seeds increased more than what was saved, and savings were mostly derived from more rational management and fewer losses due to disease and weeds.[10]

Resistance to Herbicides

One of the most studied and potentially most remunerative properties is resistance to herbicides which is conferred by inserting a gene. This is a relatively "simple" and inexpensive procedure. Its potential for profit attracted the attention of researchers of biotech companies. Plants resistant to herbicides enable higher doses of herbicide to be used, eradicating weeds more effectively. It is not by chance that corporations that market herbicides are also involved in the development and production of GMOs resistant to their products. Monsanto, the producer of *Roundup*, a widely used herbicide with active ingredient glyphosate, patented

---

[4]E. S. Opplinger, M. J. Martinka & K. A. Schimtz, *Performance of Transgenic Soybeans – Northern US*. Department of Agronomy, University of Wisconsin, Madison, WI.
[5]C. M. Benbrook, Troubled times amid commercial success for Roundup Ready soybeans: glyphosate efficacy is slipping and unstable transgene expression erodes plant defenses, Ag BioTech InfoNet Technical Paper, May, 4 (2001), available at: http://stopogm.net/sites/stopogm.net/files/TTimesBenbrook.pdf
[6]C. King, L. Purcell & E. Vories, 'Plant growth and nitrogenase activity of glyphosate-tolerant soybeans in response to foliar application', *Agron. J.* 93, 179–186 (2001).
[7]R. W. Elmore, F. W. Roeth, L. A. Nelson, C. A. Shapiro, R. N. Klein, S. Z. Knezevic & A. Martin, 'Glyphosate-resistant soybean cultivar yields compared with sister lines', *Agron. J.* 93, 408–412 (2001).
[8]J. Fernandez-Cornejo & W. D. McBride, *Adoption of Bioengineered Crops*, Economic Research Service, US Department of Agriculture, Agricultural Economics Report No. 810 (2002).
[9]C. M. Benbrook, GMO experience in the US: yields fall and pollution increases, in: *OGM: le verità sconosciute di una strategia di conquista*, Editori Riuniti, Rome (2004), pp. 59–77.
[10]M. Duffy, *Does Planting GMO Boost Farmer's Profits?*, Leopold Center for Sustainable Agriculture, Iowa State University, IA (1999).

Roundup Ready soybean. Rhone-Poulenc that produces bromoxynil "invented" bromoxynil-resistant cotton for the US market and bromoxynil-resistant tobacco for the European market. AgrEvo sells *Basta* and created *Liberty Link* plants, resistant to *Basta*. These three herbicides are defined as "total" because they kill all plants. Their extensive use over long periods eliminates weeds while their build-up in ecosystems threatens the survival of non-GM crops without resistance to these plant killers, leading to selective pressure in favour of modified seeds. This is why Rhone-Poulenc was refused a permit to sell bromoxynil in Europe: while it protects transgenic tobacco grown in the USA, it mercilessly destroys traditional tobacco.[11]

Joint development of broad-spectrum herbicides and plants modified to resist them was advertised as a milestone in the history of agriculture. Only technocrats ignorant of the basic notions of ecology could consider it such. Immunity of crops to herbicides has induced farmers to use it at any time of year, irrespective of the type of weeds and of the stage of crop growth. Obviously this has led to an increase in total herbicide use and associated pollution.[12]

Roundup is paradigmatic. Introduction of *Roundup*-resistant plants has promoted exaggerated use of the chemical, increasing costs for farmers and poisoning the soil with residues of glyphosate, which has a much longer half-life than claimed by Monsanto.[13] The half-life is the time it takes for the concentration of the substance in soil to halve. According to Monsanto, the half-life of glyphosate varies from 3 to 141 days, whereas studies by independent laboratories have found that it varies widely in relation to climate and soil, from a minimum of 55 days in Oregon forests[14] to 900 days in Swedish forests.[15] This explains the toxicity of glyphosate for many insects that protect crops against parasites,[16] and for many other microorganisms, such as mycorrhizal fungi of plant roots that fix nitrogen or confer resistance to drought and cold.[17] Glyphosate also favours proliferation of

[11] K. H. Madsen & P. Sandøe, 'Ethical reflections on herbicide-resistant crops', *Pest Manag. Sci.* 61(3), 318–325 (2004).
[12] C. Benbrook, 'Do GM crops mean less pesticide use?', *Pest. Outlook Royal Soc Chem.* October, 204–207 (2001), available at: http://www.biotech-info.net/benbrook_outlook.pdf.
[13] EPA, Office of Pesticide Programs, *Registration Eligibility Decision (RED); Glyphosphate,* Washington, DC (1993).
[14] M. Newton, 'Fate of glyphosate in an Oregon forest ecosystem', *J. Agric. Food Chem.* 32, 724–727 (1984).
[15] T. L. N. Toratenson, 'Influence of climate and adaphic factors on persistence of glyphosphate and 2,4-D in forest soils', *Eco. Environ. Safety* 18, 230–237 (1989).
[16] S. A. Hassan, 'Results of the fourth joint pesticide testing programme carried out by the IOBC/WPRS Working Group Pesticides and Beneficial Organisms', *J. Appl. Ent.* 105, 321–340 (1988).
[17] A. Towle, *Modern Biology,* Holt, Austin, Rinehart and Winston (1989), pp. 342–343.

fungal pathogens[18] and plant diseases such as anthracnosis.[19] *Roundup* is actively absorbed and accumulates in plants[20] and can be found in their edible parts. Variable concentrations of herbicide have been found in strawberries,[21] blueberries,[22] lettuce, carrots, barley[23] and even fish.[24] The long half-life of the product means that residues in soil can contaminate subsequent crops, even a year or more after its use. Not surprisingly, glyphosate is considered the third cause of occupational diseases among farmers and workers in parks and gardens of California.[25] Monsanto has the impudence to call it a friendly herbicide.

Soybeans have been found to contain glyphosate residues at levels up to 17 mg/kg.[26] Residues of glyphosate have been found in strawberries,[27] lettuce, carrots and barley planted on land previously treated with glyphosate. Glyphosate residues were found in some of these foods even when the crops were planted a year after glyphosate was applied to the soil.[28] In turn, increased concern has emerged from studies carried out on glyphosate metabolites. No maximum residue limit (MRL) has been set for glyphosate's main environmental breakdown product, AMPA (2-amino-3-(5-methyl-3-oxo-1,2-oxazol-4-yl)propanoic acid), which has been found in soybeans at high levels of up to 25 mg/kg.[29] Monsanto claims that AMPA has low toxicity to mammals and non-target organisms.[30] However,

---

[18]G. S. Johal & J. E. Rotre, 'Glyphosate hypersensitivity and phytotoxin accumulation in the incompatible bean anthracnose host–parasite interaction', *Physiol. Mol. Plant Pathol.* 32, 267–275 (1988).

[19]P. Mekwtanakam, 'Effect of certain herbicides on soil microbial populations and their influence on saprophytic growth in soil and pathogenecity of take-all fungus', *Biol. Fertil. Soils* 5, 175–180 (1987).

[20]C. Cox, 'Glyphosate, part 2: Human exposure and ecological effects', *J. Pest. Reform* 15(4), 14–20 (1995).

[21]A. J. Coxaria, 'Residues of glyphosate and its metabolite AMPA in strawberry fruit following spot and wind applications', *Can. J. Plant Sci.* 72, 1359–1366 (1992).

[22]D. N. Roy, 'Uptake and persistence of the herbicide glyphosate in fruit of wild blueberry and red raspberry', *Can. J. For. Plant* 69, 842–845 (1989).

[23]EPA, Office of Pesticide Programs, *op. cit.*

[24]L. C. Folmar, Toxicity of the herbicide glyphosate and several of its formulations to fish and aquatic invertebrates', *Arch. Environ. Contam. Toxicol.* 8, 269–277 (1979).

[25]W. S. Pease, *Preventing Pesticide-Related Illness in California Agriculture: Strategies and Priorities, Environmental Policy Program Report*, University of California, Berkeley, CA (1993).

[26]FAO, Pesticide residues in food – 2005. Report of the Joint Meeting of the FAO Panel of Experts on Pesticide Residues in Food and the Environment and the WHO Core Assessment Group on Pesticide Residues, Geneva, Switzerland, 20–29 September. FAO Plant Production and Protection Paper (2005), 183, 7.

[27]A. J. Cessna & N. P. Cain, 'Residues of glyphosate and its metabolite AMPA in strawberry fruit following spot and wiper applications', *Can. J. Plant Sci.* 72, 1359–1365 (1992).

[28]US EPA, *op. cit.*

[29]H. Sandermann, 'Plant biotechnology: ecological case studies on herbicide resistance', *Trends Plant Sci.* 11, 324–328 (2006).

[30]Monsanto. Backgrounder: Glyphosate and environmental fate studies. Monsanto (April 2005).

recent research testing the effects of *Roundup* formulations found that both AMPA and the *Roundup* adjuvant polyethoxylated tallow amine (POEA) kill human cells at extremely low concentrations.[31] A study found that AMPA causes DNA damage in cells.[32] POEA is about 30 times more toxic to fish than glyphosate.[33]

Intensive use of total herbicides does not exclude the emergence of weeds resistant to the herbicide, despite unconvincing denials from the industry.[34,35] This has already occurred by two distinct mechanisms, well known to microbiologists and pharmacologists: repeated use of a single herbicide can give rise to herbicide-resistant weeds that supplant traditional weeds through selective pressure; secondly, horizontal gene transfer can enable plants related to or growing near transgenic crops to acquire genes conferring resistance to the herbicide.

Gene Transfer

GM plants have a twenty-fold probability of horizontal transmission of the introduced gene with respect to plants acquiring the gene segment by simple cross- or spontaneous mutation.[36] In turn, the new gene can be transmitted to progeny (vertical transmission); this is true for plants of the same or different species. Examples of spontaneous hybridization between food plants and wild plants have been documented in 12 out of 13 cases of major edible varieties (sugar beet, canola, potato) and for many other species, the progeny of which not only incorporates the acquired genetic characters but has a reproductive capacity similar to that of the parent plants.[37] Transfer of genes conferring resistance to herbicide has been documented in certain varieties of *Brassica*, where the phenomenon occurs quite

---

[31]N. Benachour & G.-E. Séralini, 'Glyphosate formulations induce apoptosis and necrosis in human umbilical, embryonic, and placental cells', *Chem. Res. Toxicol.* 22, 97–105 (2009).
[32]F. Mañas, L. Peralta, J. Raviolo, H. Garcia Ovando, A. Weyers, L. Ugnia, M. Gonzalez Cid, I. Larripa & N. Gorla, 'Genotoxicity of AMPA, the environmental metabolite of glyphosate, assessed by the Comet assay and cytogenetic tests', *Ecotoxicol. Environ. Safety* 72, 834–837 (2009).
[33]J. A. Servizi, R. W. Gordon & D. W. Martens, 'Acute toxicity of Garlon 4 and Roundup herbicides to salmon, *Daphnia* and trout', *Bull. Environ. Contam. Toxicol.* 39, 15–22 (1987).
[34]A. J. Gray & A. F. Raybould, 'Crop genetics: reducing transgene escape routes', *Nature* 392, 653–654 (1998).
[35]R. Stanley & U. Baumann, 'Resistance to herbicide glyphosate', *Nature* 395, 25–26 (1998).
[36]J. Bergelson, C. B. Purrington & G. Wichmann, 'Promiscuity in transgenic plants', *Nature* 395, 25 (1998).
[37]R. N. Mack, D. Simberloff, W. M. Lonsdale, H. Evans, M. Clout & F. Bazzaz, 'Biotic invasions: causes, epidemiology, global consequences, and control', *Issues Ecol.* 5, 1–22 (2000).

quickly,[38] though it only affects a small fraction of the plants.[39] The probability of transfer varies in relation to the space separating the two plant species, ranging from 24% for modified potatoes cultivated in rows alternating with traditional tubers to 0.00001% for plants growing 50–400 m apart. Although this risk is low,[40] it is not zero and also depends on climatic conditions and the abundance of pollinating insects.[41] Escape of crop transgenes into wild relatives via hybridization or virus-infected transgenic plants producing more virulent (pathogenic) genotypes with recombined mRNA of the transgene should not be dismissed lightly.[42]

The risk of invasion by new species with "modified" characters is a result of interaction between a species (hybrid) and the ecosystem; it cannot reasonably be assessed by analysing the characteristics of a species (parental) alone without considering the environment and specific agricultural conditions (rotation, monoculture, size of cultivated area, and so forth).[43]

In both the above cases, the risk of relatively fast emergence of herbicide-resistant plants and weeds is very high.[44] In this case it is necessary to change the type of seed and the herbicide and begin the vicious circle from scratch; the only winners are the agrochemical companies. Some scientists have realized the risks and documented the possibility of gene transfer to wild and non-modified species, underlining the need for protection and confinement (such as sowing only male-sterile cultivars of potatoes).[45,46]

The marketing of plants resistant to a cocktail of different herbicides would only worsen the problem, favouring emergence of weeds with multiple resistance, a disastrous eventuality. In such cases the only approach would be the optimistic faith shown by agrochemical multinationals, that instead of being concerned about

[38]T. R. Mikkelsen, B. Andersen & R. B. Joergensen, 'The risk of crop transgene spread', *Nature* 380, 31–34 (1996).

[39]OTA US Congress, Office of Technology Assessment. *Harmful Non-Indigenous Species in the United States*, OTA-F-565, Washington, DC (1993).

[40]A.-M. Chevre, F. Eber, A. Baranger & M. Renard, 'Gene flow from transgenic crop', *Nature* 389, 924 (1997).

[41]V. S. Luna, M. J. Figueroa, M. R. Baltazar, L. R. Gomez, R. Towsend & J. B. Schoper, 'Maize pollen longevity and distance isolation requirements for effective pollen control', *Crop Sci.* 41, 1551–1557 (2001).

[42]M. J. Scriber, 'Bt or not Bt: is that the question?', *Proc. Natl Acad. Sci. U. S. A.* 98(22), 12328–12330 (2001).

[43]H. Torgersen, G. Soja, I. Janssen & H. Gaugitsch, Risk assessment of conventional plants in analogy to transgenic plants', *ESPR – Environ. Sci. Pollut. Res.* 5(2), 89–93 (1998).

[44]H. Gaugitsch, 'Experience with environmental issues in GM crop production and the likely future scenarios', *Toxicol. Lett.* 127, 351–357 (2002).

[45]C. Celis, M. Scurrah, S. Cowgill, S. Chumbiauca, J. Green, J. Franco, G. Main, D. Kiezebrink, R. G. F. Visser & H. J. Atkinson, 'Environmental biosafety and transgenic potato in a centre of diversity for this crop', *Nature* 432, 222–225 (2004).

[46]J. K. Hancock, 'A framework for assessing the risk of transgenic crops', *Bioscience* 53, 512–519 (2003).

tomorrow, prefer to tackle problems one at a time as they arise. It is worrying that the *Nuffield Council of Bioethics* considered such risks "insufficient reason" for limiting "the introduction of GMO into developing countries"[47] and one may well ask what type of demonstration would suffice to solicit a critical reappraisal of GMOs. Available data does not suggest a rosy future. A study group of Charles Sturt University identified a weed, *Lolium perennis*, which had become insensitive to *Roundup*, tolerating five times the maximum recommended dose.[48] Aware of the progressive emergence of the phenomenon of resistance, Monsanto first tried to deny the evidence of horizontal gene transfer,[49] and was then forced to negotiate with the authorities of many nations to raise the limit of herbicide residues, bringing it to 20 mg/kg of dry product. The prospect offered by intensive use of *Roundup*-resistant GMO crops is therefore a rising contaminant load due to increasing use of pesticides. This will pollute soil and water, affecting the fertility and reproduction of many useful insects. Economic losses due to proliferation of resistant weeds that compete with crops for nutrients and water impose escalating herbicide use, estimated at 10–15% of crop value. This is not a comforting figure and contrasts with the expectations promoted by company advertising. One could object that higher yields and better quality make it worthwhile, but even this is untrue. Many new crops, for example GM cotton in 1997, proved a complete flop and caused millions of dollars of damage. Others, such as GM soybeans, gave poor yields and proved to be of inferior quality due to loss of the bioflavonoids, genisteine and daidzeine,[50] which are the biopharmacological added value of the plant. Soy is widely used in dietology to treat menopausal symptoms and to prevent hormone-dependent tumours by virtue of its high plant oestrogen content. Monsanto GM soya with 20% fewer bioflavonoids is of less medical interest and less economically attractive. So where is the advantage?

Uncle Scrooge's Sweater

According to a Mickey Mouse comic by Walt Disney,[51] Uncle Scrooge and his nephews, Hewey, Dewey and Lewey, discover a miraculous type of wool that keeps

---

[47] Nuffield Bioethics Committee. *The Use of Genetically Modified Crops in Developing Countries*, Nuffield Bioethics Committee, London (2004).

[48] L. Landbo & R. B. Jorgensen, '*Brassica campestris* and its hybrids with *B. napus*: implications for risk assessments of transgenic oilseed rape', *Euphytica* 97, 209–216 (1997).

[49] Monsanto researchers unwillingly admitted that in some cases transfer of resistance genes to noxious weeds was possible (L. D. Bradshaw, S. R. Padgette, S. L. Kimball & B. H. Wells, 'Perspectives on glyphosate resistance', *Weed Technol.* 11, 189–198 (1997)).

[50] M. A. Lappé, E. B. Bailey, C. Childress & K. D. R. Setchell, 'Alterations in clinically important phytooestrogens in genetically modified herbicide-tolerant soybeans', *J. Med. Food* 1, 241–245 (1999).

[51] W. Disney, *Paperino e il vello refrigerante* in. I Classici Disney, March 2001, n. 292.

the wearer cool. The wool is used by the people of a remote highland to make sweaters that paradoxically protect from heat and insolation by local refrigeration of the microenvironment. Scrooge immediately realizes the commercial potential. He wastes no time in getting his hands on the "biotechnological" treasure and exploits it on an industrial scale. Hundreds of thousands of buyers from all over the world rush to purchase his sweaters of magic wool. People are happy because they can finally lie in the sun without becoming sunburnt or suffering heatstroke, and can live comfortably in the desert. Scrooge grows richer and richer. Then one day, an epidemic of colds, rheumatism and flu breaks out. The people of Scroogeville followed by the rest of the world all line up at the doctor's surgery. The diagnosis is the same for all: colds and flu. The cause soon emerges: the mass of refrigerating sweaters has not only modified the microclimate but also the climate of the planet, and the people of the earth are exposed to an unusual wave of cold in the middle of summer, with accompanying winter illnesses. Scrooge weeps bitter tears as he pays reparations and withdraws the garments from the market in order to restore biological and meteorological equilibrium.

Like all fables, this story was written for children in order to be understood by adults, to whom it suggests a lesson rich in implications. The thoughtless actions of humans that upset the equilibrium of countless ecosystems can have sudden repercussions, especially when mixing of different ecological niches is favoured: that wool, for example, should never have left the highlands of its origin and been distributed throughout the world, as is occurring with "modified" seeds. If the reader thinks these are fables, he may change his mind after reading the next section.

Can Microbes Change Climate?

It is not a remote or even improbable possibility that radical changes induced by humans on a planetary scale upset biological and climatic equilibria. An interesting example is offered by one of the first GM organisms, the release of which was authorized by the US government in the early 1980s. Researchers of the University of California modified *Pseudomonas syringae*, a bacterium common in temperate regions. The only reason researchers were interested in it was that it incorporates ice crystals in its cytoplasm. Acting like a miniature refrigerator, the microbe has always been a pest for crops. Attaching to plant leaves, it creates a layer of ice that "burns" the plant. Molecular biologists decided to create a variant of *Pseudomonas syringae* without the gene sequence that codes the information necessary to form ice crystals. They named the new strain *ice minus*. They thought that by spraying large quantities of the new bacterium on fields, they could exercise selective pressure and replace the original *Pseudomonas* pest. The advantages of the operation were evident, or at least seemed to be at first. The damage and the costs came later, as in the case of Scrooge's sweaters.

According to Eugene Odum, an esteemed figure of American ecology, professor at the University of Georgia, the bacterium has important functions and an irreplaceable role:

> It seems that the lipoprotein coats of this and other species of bacteria found on plants and in detritus when shed and wafted up into the clouds form ideal nuclei for ice formation that is absolutely necessary for rain to fall [...] If *Pseudomonas syringae* does indeed have a beneficial role in enhancing rainfall, then the ecologist's concern about possible secondary or indirect effects of releases of GM organisms is vindicated.[52]

This concept was sustained by studies of the *National Oceanic and Atmospheric Administration Laboratory* at Boulder (California) and confirmed by Prof. Lindow who was charged by the University of California to investigate the actual and potential repercussions of *ice minus* on climate.[53] For once, the organization that invented the new microbe did not deny the evidence and recognized that natural non-modified bacteria can be an important factor in the processes of precipitation.[54] Dangers of this kind were not imagined and could not be predicted from field tests on single small plots of land. These tests were also unsuitable for assessing the impact modification of the genome of an ecosystem component that could have on overall ecological equilibrium.

The most acute comment on this story comes from one of the fathers of modern genetic engineering, Renato Dulbecco, who observed

> a basic contradiction in these experiments. Let us consider the *ice minus* bacterium. If it is innocuous, as they say, that is, non invasive, it will soon be overcome by the pre-existing species and will disappear. In order to be truly effective in time, it would have to prevail over the other bacteria, in which case it would be dangerous. In other words, either this technology is not dangerous but useless, or it is useful but dangerous.[55]

Deaf to the warnings of scientists (not only environmentalists), the US Court of Appeal authorized marketing of *ice minus*, since the "evidence" produced in field tests was negative. The Court did not ask whether such a radical act could have risks and repercussions on ecosystems. In authorizing the introduction of a modified organism, no attempt was made to fully understand the complex biological system involved.

---

[52]E. P. Odum, 'Biotechnology and the biosphere', *Science* 229(4720), 1338 (1985).
[53]S. F. Lindow, 'Methods of preventing frost injury caused by epiphytic ice-nucleation active bacteria', *Plant Dis.* 3, 327–337 (1983).
[54]S. E. Lindow, *ibid.*, p. 332.
[55]R. Dulbecco, *Ingegneri della vita*, Sperling & Kupfer, Milan (1988), p. 121.

The existence of *Pseudomonas syringae* is not a freak of nature or due to chance. Any biologist who thinks so should change his profession. The first task of a scientist worthy of the name is to understand the role and function that a microbe with such unique properties could have, not to modify something he does not understand. To our knowledge, no one tried to do this, at least not before human ineptitude upset an equilibrium in which that insignificant microbe was a fulcrum. Here lies the difference with respect to the example of Uncle Scrooge: withdrawal of the refrigerating sweaters enabled the lost equilibrium to be restored. In the case of genetic modifications, the transformations affect the progeny and are imposed furtively on the natural varieties, with irreversible results. Proceeding blindly in this manner, nothing and no one can reassure us that lost equilibrium can be restored. Unfortunately, there is no shortage of examples.

Bizarre Bacteria

Drought and organic impoverishment of soils are the two main causes that are making large areas of our planet sterile. Soil fertility should be increased through fixation of nitrogen, which is normally carried out by certain bacteria. Hopes were raised by modification of *Bradyrhyzobium japonica*,[56] a microorganism that intercepts atmospheric nitrogen and stores it in its cytoplasm. *Biotechnica International*, the company that created the mutant version of the bacterium, made an agreement with an experimental station of the US Department of Agriculture to field test its efficacy in Louisiana. Soybeans were "contaminated" with the bacteria, sown and once they had grown were burned to be ploughed into the soil. The nitrogen content of the area, about two hectares, increased as expected. What was not expected was the disappearance of the natural strain of the microbe that was somehow destroyed by its GM counterpart. Such results are disagreeable not only to ecologists but also to the whole scientific community: if a newly introduced genetic variant reduces biodiversity so strongly, it can have harmful effects that lead to the disappearance of other forms of life. The consequences are unpredictable and negative by definition. Nature has a complex system of weights and counterweights that work together to protect the environment from sudden and irreversible imbalances: any perturbation is counteracted by a series of counter-reactions that maintain ecosystem homeostasis.

An example from biology textbooks is the relation between population flows of silver foxes and lichens in the Siberian tundra. The lichen is a preferred food of foxes. If the lichen population grows, the foxes have more food and their population grows until the excess of foxes leads to a relative scarcity of lichens. The reduced presence of lichens means that less food is available for foxes and their population decreases to a level at which the lichen population can recover and the cycle begins

---

[56] J. Cairns, 'Establishing environmental hazards of genetically engineered microorganisms', *Rev. Environ. Contam. Toxicol.* 124, 19–23 (1992).

again. If foxes or lichens were eliminated, the whole ecosystem would suffer. The same occurs when a new biological factor is introduced into an environment, without any feedback to control it.

Tragic confirmation of this scientific truth is offered by GM *Klebsiella* introduced to produce ethanol from agricultural waste. Initially conceived to obtain cheap motor fuel, modified *Klebsiella* strains nourished hopes that were met with disappointment and tragedy. The GM bacteria dramatically altered the soil nutrient substrate and inhibited the growth of roots and sprouts of wheat; the microbe killed symbiotic fungi, promoting the development of nematodes[57] and other protozoan parasites, affecting the micronutrient composition of surface soil and harming plant growth.[58] Although survival of the transgenic microbe in soil was longer than expected, it was not eternal and it was possible to repair the damage. Next time, undoing the effects of a microbe with unknown properties, light-heartedly released into the environment, may be much more difficult.

The examples do not finish here. Another "brilliant" idea was to modify certain bacteria so that they could metabolize the residues of a persistent herbicide, 2,4-D. The first experiments showed good efficacy of the GMO but nobody thought to look for unexpected effects. Unfortunately, these occurred at the time of application. Once nebulized on the soil, the bacteria digested the herbicide residues, producing an extremely toxic metabolite, 2,4-Dcp, which accumulated in the soil, where in a few weeks it killed many species of fungi essential for crop growth.[59]

The fate of the monarch butterfly *Danaus plexippus*[60] is another unhappy story. Scientists from Cornell University under Prof. Losey showed that its larvae were destroyed by pollen of maize in which the *Bt*-gene had been inserted. The pollen killed about half the insects in 3–4 days. Those that survived showed malformations.

"We Arrived 20 Million Acres of Bt-Maize Too Late"

The data produced by Cornell University raised great controversy. New research, parallel to clarifications and new data produced by Losey's group, was carried out and published. Many of these studies did not spare criticism of Losey's findings but they confirmed the alarm raised by the first study. The methodological defect

---

[57] M. T. Holmes, 'Effects of *Klebsiella panticola* SDF20 on soil biota and wheat growth in sadly soil', *Appl. Soil Ecol.* 11, 67–74 (1999).
[58] M. T. Holmes & E. R. Ingham, *Bull. Ecol. Am.* (Abs.) 7S, 2 (1994).
[59] J. D. Doyle, G. Stotzky, G. McClung & C. W. Hendricks, 'Effects of genetically engineered microorganisms on microbial populations and processes in natural habitats', *Adv. Appl. Microbiol.* 40, 237–244 (1995).
[60] J. E. Losey, L. S. Rayor & M. E. Carter, 'Transgenic pollen harms monarch larvae', *Nature* 399, 214 (1999).

generally invoked was that Losey did not specify the doses of pollen to which the larvae were exposed. The study[61] most critical of the original findings showed that Losey's first results[62] may have been affected by contaminants and certain transgenic variants of corn. The butterfly has differentiated sensitivity to the toxic effects of the protein expressed by Bt-corn, depending on its stage of development and the prevalence of the various proteins expressed by the transgene (Cry1Ab, Cry1Ac, Cry9C, Cry1F), which vary in different hybrids. Cereal variants that express Bt-correlated Cry9c and Cry1F do not seem to have substantial effects on butterfly viability. However, in the words of the authors:

> Bioassays of purified *Bt* toxins in artificial diet indicate that Cry9C and Cry1F proteins are relatively nontoxic to monarch first instars, whereas first instars are sensitive to Cry1Ab and Cry1Ac proteins [...] Results from pollen bioassays suggest that monarch sensitivity to Cry proteins varies depending on the amount of sample contamination with nonpollen material and the specific transgenic event [...] The only transgenic corn pollen that consistently affected monarch larvae, even at low levels, was from event 176 hybrids, which use a pollen-specific promoter for expression of the *cry1Ab* gene.[63]

Although the above study limited and specified the setting of the first study, it did not deny that Bt-maize can affect the survival of butterfly larvae. Indeed, it clarified that one of the varieties in question (hybrid 176) reduced larval viability substantially even at low protein doses.

Similar results were obtained by the research groups of Carrière[64] and Wraight.[65] The latter published two papers on the monarch butterfly and *Papilio polixenes*, which while they did not confirm the previous results on the monarch, nevertheless

---

[61]D. E. Stanley-Horn, G. P. Dively, R. L. Hellmich, H. R. Mattila, M. K. Sears, R. Rose, L. C. H. Jesse, J. E. Losey, J. J. Obrycki & L. Lewis, 'Monarch larvae sensitivity to *Bacillus thuringiensis*-purified proteins and pollen, *Proc. Natl Acad. Sci. U. S. A.* 98, 11925–11936 (2001).

[62]Losey replied that larvae on leaves of *A. syriaca* dusted with Bt-maize pollen showed significantly higher mortality after 48 h (20 + 3%) than larvae on leaves not contaminated with pollen (3 + 3%). At 120 h larvae of *D. plexippus* exposed to 135 grains of transgenic pollen/cm² ranged from 37% to 70%. The effects of Bt-pollen were detected up to 10 m from the edges of the transgenic field. The ecological effects of plants modified genetically to express insecticide properties must be very carefully evaluated before they are grown on large areas (L. C. H. Jesse, J. J. Obrycki & J. E. Losey, 'Field deposition of *Bt* transgenic corn pollen: lethal effects on the monarch butterfly', *Oecologia* 125(2), 241–248 (2000)).

[63]D. E. Stanley-Horn, G. P. Dively, R. L. Hellmich, H. R. Mattila, M. K. Sears, R. Rose, L. C. H. Jesse, J. E. Losey, J. J. Obrycki & L. Lewis, *op. cit.*

[64]Y. Carrière, C. Ellers-Kirk, M. Sisterson, L. Antilla, M. Whitlow, T. J. Dennehy & B. E. Tabashnik, 'Long-term regional suppression of pink bollworm by *Bacillus thuringiensis* cotton', *Proc. Natl Acad. Sci. U. S. A.* 100, 1519–1523 (2003).

[65]C. L. Wraight, A. R. Zangerl, M. J. Carroll & M. R. Berenbaum, 'Absence of toxicity of Bacillus thuringiensis pollen to black swallowtails under field conditions', *Proc. Natl Acad. Sci. U. S. A.* 97, 7700–7703 (2000).

showed that *P. polixenes* was sensitive to pollen from *Novartis* hybrid 176, a modified maize in which Bt-expression is about 40 times greater than in the variety patented by Monsanto (MON810).

Again, in the words of the authors:

> However, despite five rainfall events that removed much of the pollen from the leaves of their host plants during the experiment, we observed a significant reduction in growth rates of black swallowtail larvae that was likely caused by pollen exposure. These results suggest that *Bt* corn incorporating event 176 can have adverse sublethal effects on black swallowtails in the field and underscore the importance of event selection in reducing environmental impacts of transgenic plants.[66]

Even more interesting is the work of Stanley-Horn, which confirmed the toxicity of hybrid 176 of Bt-maize, linking it directly to intense expression of toxin, significantly greater than by other variants such as MON810, while pointing out that further research was needed to better understand the impact of Bt-pollen having relatively modest toxicity on the monarch butterfly population. Not only concentration but also duration of exposure must be considered in calculating environmental dose.[67] A recent study by Stanley-Horn's group underlined again that the toxicity of GM maize due to expression of one or more Cry proteins depends on various factors: the nature of the hybrid, the quantity of protein expressed, the degree of exposure (concentration × duration) and, last but not least, the stage of development of the butterfly.[68] The concluding comment by David Pimentel and Peter Raven in the prestigious *Proceedings of the National Academy of Sciences USA* (PNAS)[69] was that the toxicity of Bt-maize varies in relation to the type of *Lepidoptera* considered and in relation to the different content of Bt-proteins expressed, calling for further research to limit or prevent damage to non-target butterflies and other beneficial insects.

Beyond the controversy surrounding the unfortunate butterfly, the *Institut National pour la Récherche Agricole* (INRA) presented an instructive comment at a workshop at the University of Berkeley:

> Until then, the official risk assessment had managed to avoid considering the effect of the Bt toxin on non-target insects. In this

---

[66]R. A. Zangerl, D. McKenna, C. L. Wraight, M. Carroll, P. Ficarello, R. Warner & M. R. Berenbaum, 'Effects of exposure to event 176 *Bacillus thuringiensis* corn pollen on monarch and black swallowtail caterpillars under field conditions', *Proc. Natl Acad. Sci. U. S. A.* 98(21), 11908–11912 (2001).
[67]D. E. Stanley-Horn, G. P. Dively, R. L. Hellmich, H. R. Mattila, M. K. Sears, R. Rose, L. C. H. Jesse, J. E. Losey, J. J. Obrycki & L. Lewis, *op. cit.*, pp. 11931–11936.
[68]M. K. Sears, R. L. Hellmich, D. E. Stanley-Horn, K. S. Oberhauser, J. M. Pleasants, H. R. Mattila, B. D. Siegfried & G. P. Dively, 'Impact of Bt-corn pollen on monarch butterfly populations: a risk assessment', *Proc. Natl Acad. Sci. U. S. A.* 98(21), 11937–11942 (2001).
[69]D. S. Pimentel & P. H. Raven, 'Bt corn pollen impacts on nontarget Lepidoptera: assessment of effects in nature', *Proc. Natl Acad. Sci. U. S. A.* 97(15), 8198–8199 (2000).

context, the criticism about the methodological limitations of the study reinforced its message. [...]This study [...] raised the question of why such experiments were not performed earlier [...] *Why is it that this study was not done before the approval of Bt corn? This is 20 million acres of Bt corn too late. This should serve as a warning that there are more unpleasant surprises ahead.*[70]

Insect-Resistant Plants

Molecular biologists consider that an alternative to massive use of pesticides is to insert genes into plants to give them some form of "resistance" to insects, through production of substances toxic for the insect or compounds that make the plant less palatable to natural predators. Actually, the researchers gleaned this "idea" from observations in nature. A classic example is the potato, *Solanum tuberosus*, rich in solanines, alkaloids with a sterol base similar to many steroid hormones. Solanines and alkaloids in general are not essential for plant life, being metabolic by-products that play a defensive role. Wild solanaceae are especially rich in solanines and are more resistant to parasites such as *Fusarium* and insects such as *Dorifera*. Animals avoid plants having high alkaloid content, and indeed solanine was once used as an agricultural insecticide. Plants developed alkaloids through a long process of climatization and coevolution that preserved the equilibrium between insects, plants and animals in a given ecosystem. Plants developed a defence strategy by elaborating variable quantities of a chemical weapon. Since solanine is cardiotoxic for humans, potato strains with lower and lower alkaloid contents have been selected, making the potato increasingly susceptible to predators, ranging from *Dorifera* and mole crickets to fungi. Resort was therefore made to pesticides to protect the yield. An attempt to avoid excessive use of pesticides led to the development of *Bacillus thuringiensis* (Bt) in the 1960s, a "biological" solution to the problem.

Bacillus thuringiensis is an aerobic microbe that produces different classes of toxins in the form of tiny crystals of protein that build up in its cytoplasm, especially during sporulation, the process by which it reproduces through budding of spores. If absorbed by an insect, the toxins induce progressive paralysis of the digestive system, leading to death. *Bacillus thuringiensis* produces at least 50 different proteins, selectively specific for certain predators, for each of which the respective Cry gene has been isolated and cloned. The toxins are sprayed on crops

---

[70]Agenda setting and controversies: a comparative approach to the case of GMOs in France and the United States, available at: https://docs.google.com/viewer?a=v&q=cache :OxydZFfsMXYJ:citeseerx.ist.psu.edu/viewdoc/download?doi%3D10.1.1.199.1333%26 rep%3Drep1%26type%3Dpdf+Agenda+setting+and+controversies:+a+comparative+ approach+to+the+case+of+GMOs+in+France+and+ the+United+States,&hl=it&gl= it&pid=bl&srcid=ADGEEShiUHc54q0GgpOodDmu7hSWaYpTzk_REVwuZElbuD0UGu AupGDIWnu0ZlflaeaH6bnUzXgTUTP6c0IYssHjNyG7-g8XE8QA2WSGwZ2BenyC4Kqbat_ oeMQc9NqxseCz2LyTp4bh&sig=AHlEtbRunXjF9kFhbincAf6HQbld5JYoig&pli=1, Rapport INRA, 8–9 June (2001), pp. 17–18.

and are substantially risk-free, being quickly broken down by light. They are not toxic for birds or mammals but are highly specific for the insect to be fought. The US EPA recognized Bt as the least environmentally harmful pesticide and among the most effective and safe, as early as 1961. The product was an enormous commercial success, production rising from about 300 tons in 1985 to 1500 tons in 1990. Should parasites develop resistance to Bt, it would be necessary to resort to chemical pesticides which are much more costly, polluting and dangerous.

The "scientists" of Plant Genetic Systems Inc. had the idea of transferring the Cry gene to different types of plants, so that they would produce toxic proteins themselves, all year round. The project was launched with great publicity as a demonstration of how technology worked side by side with nature in the struggle against predatory insects, without the need for harmful, costly synthetic chemicals. The idea was only apparently good. Indeed, transgenic plants expressing the Cry gene may continuously synthesize excessive quantities of the toxin which was previously only applied in the case of "need" and in certain seasons. The toxins sprayed on crops were specific for the insects to be eliminated, whereas the modified plants randomly expressed only one of the fifty possible protein variants transcripted from the introduced gene. Spraying of Bt-toxin treated the surface of the plant and did not penetrate into plant tissues, and it was administered in the form of prototoxin. In other words, the biological pesticide was sprayed in a metabolically inert form that only became active if ingested by an insect. In the case of plants bearing the Cry gene, the toxin was synthesized in active form and as such acted on any form of life with which it had contact. The toxin produced directly by the plant tended to remain longer in the soil where its toxicity persisted three times longer than that recorded for the "biological" counterpart.[71] Despite this evidence, the implications of which any molecular biologist should have recognized at once, the fad of transforming crops by inserting the Bt-gene spread like wildfire. The first plant was tobacco (1986), followed by maize, cotton and potatoes. In 1997, 21 out of almost 80 GM crops synthesized Bt-toxin. In the USA alone, Bt-modified crops occupied more than 1,500,000 ha: 800,000 of maize and 700,000 of cotton by Monsanto, 200,000 of maize by Novartis, 20,000 of potatoes by Monsanto.[72] With regard to cotton, 15% of the cultivated area was already allotted to transgenic varieties. A lucrative future was anticipated by champions of GMOs, but their triumphant march was soon arrested. As biologists have always known, pests develop resistance, and uncontrolled spread of Bt-crops induced resistance much sooner than expected. The first tests indicated that this possibility was remote, a mere detail in the overall picture of transgenic strategy. But the devil hides in details ....

---

[71]R. A. Steimbrecher, 'From gene to gene revolution. The environmental risks of genetically engineered crops', *Ecologist*, November/December, 275–277 (1996).
[72]A. F. Krattiger, 'Insect resistance in crops: a case study of *Bacillus thuringiensis* (Bt) and its transfer to developing countries', *ISAA Briefs* 2 (1997).

The Incredible Story of Bt-Toxins

Resistance to various substances used to fight insect pests is a major problem in agriculture. With intensive development of monocultures and the consequent massive use of insecticides, the species of mites and insects resistant to insecticides were soon more than 50.[73] Though specific and used with care, not even Bt-toxin was an exception to the rule: the selective pressure of a pesticide used against a broad population (in this case, insects) sooner or later leads to selection of a naturally "resistant" strain that proliferates despite the pesticide. The emergence of Bt-resistant insects was recorded by EPA in 1981 and since then confirmation and new reports have escalated,[74] contradicting the claim that only insects "selected" in the laboratory showed this feature.[75] The situation precipitated with extensive cultivation of Bt-crops, where selective pressure on a wide variety of insects that have relations with the plant is high. Unlike Bt-products applied by spraying, Bt-modified plants secrete great quantities of toxin in active form throughout the year.[76] Out of eight species of insects studied in early 1992, at least one, *Plutella xylostella*, better known as the diamondback moth, showed high resistance to Bt. The study showed what has always been known, namely that biological and "alternative" pesticides can give rise to the same problem as synthetic ones, especially if used inappropriately.[77]

However, it was only in 1997 that producers of modified plants felt the full impact of their actions. Direct estimation of the frequency of insects resistant to Bt-endotoxin showed that one in 350 specimens of tobacco budworm, *Heliotis virescens*, a major parasite of cotton, was Bt-resistant. This percentage was about 10 times that predicted by theoretical models and signalled "high risk of rapid adaptation".[78] The high incidence of insects immune to Bt was a warning that cotton carrying the Cry gene could only last about 10 years before being again at the mercy of the tobacco budworm. Study of other parasites further reduced any hopes, indicating that such crops could not last longer than 3–4 years, provided "reserve" areas, 4–5% of the area cultivated with cotton were created where non-transgenic cotton was grown, to enable proliferation of non-Bt-resistant

---

[73]W. H. McGaughey & M. E. Whalon, 'Managing insect resistance to *Bacillus thuringiensis* toxins', *Science* 258, 1451–1453 (1992).

[74]'Noting that cabbage loopers and diamond-back moth caterpillars have developed resistance to Bt', *Fed. Reg.* 14445, 27 February (1981).

[75]B. E. Tabashnik, Y. Carriere, T. J. Dennehy, S. Morin, M. S. Sisterson, R. T. Roush, A. M. Shelton & J. Z. Zhao, 'Insect resistance to transgenic Bt crops: lessons from the laboratory and field', *J. Econ. Entomol.* 96, 1031–1038 (2003).

[76]M. E. Whalon & D. L. Norris, 'Resistance management for transgenic *Bacillus thuringiensis* plants', *Biotechnol. Dev. Monitor* 29, 812 (1996).

[77]B. E. Tabashnik BE, 'Evolution of resistance to *Bacillus thuringiensis*', *Annu. Rev. Entomol.* 39, 47–79 (1994).

[78]F. Gould, 'Sustainability of transgenic insecticidal cultivars: integrating pest genetics and ecology', *Annu. Rev. Entomol.* 43, 701–726 (1998).

insects.[79] Actually, this result was far from unexpected. Already during the first field tests, it was seen that only 80% of the parasites were killed by plants transfected with the gene coding for Bt-endotoxin. A 20% survival rate ensures that resistant strains will give rise to selection of a race of super-insects that can rapidly replace sensitive strains and prevail.

This story is the umpteenth demonstration of the superficiality and ignorance with which molecular biologists tackle complex topics like the equilibrium of species in ecosystems. Prof. Fred Gould of North Carolina University wrote that 80% mortality is exactly what scientists aim at when they want to produce resistant insects.[80]

After much prevarication and denial of the evidence, the US government and the producing companies began to realize the risk:[81,82] the new transgenic varieties that challenged time and insects would soon be useless because of the very manipulation intended to give them an evolutionary advantage. Resistance to Bt-biopesticides manifested more than 10 years ago; since then, the eight main insects species (Colorado potato beetle, diamondback moth, tobacco budworm, etc.) have all developed substantial immunity to Bt-toxin.[83] Measures were therefore necessary. The EPA recommended the creation of "reserve zones" for the moth, at least 30% of the total area cultivated with Bt-maize. If other insecticides were used, the percentage increased to 40%: farmers found themselves in the paradoxical and tragicomic situation of having to cultivate almost half of their fields with GM and half with traditional maize. The story does not end there: the whole strategy was susceptible to failure due to transfer of the genes of resistance of the modified species to the non-modified ones: thus contamination of the refuge areas would amplify the phenomenon of resistance, extending it to various species of predatory

---

[79]F. Gould, A. Anderson, A. Jones, S. Sumerford, D. G. Heckel, J. Lopez, S. Micinski, R. Leonard & M. Laster, 'Initial frequency of alleles for resistance to *Bacillus Thuringiensis* toxins in field populations of *Heliothis virescens*', *Proc. Natl Acad. Sci. U. S. A.* 94, 3519–3523 (1997). The need to have areas assigned to non-GM crops enables the survival of insects not subjected to the selective pressure of Bt-endotoxin that can compete with mutants resistant to Bt. This measure would counteract dominance of the resistant species but is obviously a poor palliative for the emergency and is unlikely to work. The gene conferring resistance to Bt-toxin is usually conserved for more than 100 generations without exposure to Bt. This characteristic of the gene indicates that any reduction in selective pressure that it is intended to achieve through non-transgenic refuges is likely to have little effect. (B. E. Tabashnik, Y.-B. Liu, N. Finson, L. Masson & D. G. Heckel, 'One gene in diamondback moth confers resistance to four *Bacillus thuringiensis* toxins', *Proc. Natl Acad. Sci. U. S. A.* 94, 1640 (1997)).
[80]Quoted in J. L. Fox, 'Cotton infestations renew resistance concern', *Nat. Biotechnol.* 34, 1070 (1996).
[81]Available at: www.epa.gov/pesticides/biopesticides/pips/bt_brad.htm.
[82]A. A. Snow, 'Transgenic crops – why gene flow matters', *Nat. Biotechnol.* 20, 542 (2002).
[83]J. Rissler & M. Mellon, *The Ecological Risks of Engineered Crops*, MIT Press, Cambridge, MA (1996), p. 43.

insects,[84,85] and making the reserve zones completely useless.[86,87] Gene transfer from Bt-maize to non-modified maize could accelerate the processes of acquisition of resistance by insects in two ways: (a) the presence of Bt-toxin in refuge zones would kill susceptible larvae and help resistant insects to emerge; (b) production of toxin in non-Bt areas below recommended standards, while killing susceptible larvae, would enable heterozygotes to survive, increasing the functional dominance of carriers of resistance, which would therefore evolve faster. After culpable under-estimation by the first studies,[88,89] this scenario was confirmed experimentally by many renowned researchers:[90]

> The highest Bt toxin concentration in pooled kernels of non-Bt maize plants was 45% of the mean concentration in kernels from adjacent Bt maize plants. Most previous work on gene flow from transgenic crops has emphasized potential effects of transgene movement on wild relatives of crops, landraces, and organic plantings, whereas implications for pest resistance have been largely ignored. Variable Bt-toxin production in seeds of refuge plants undermines the high-dose/refuge strategy and could accelerate pest resistance to Bt crops. Thus, guidelines should be revised to reduce gene flow between Bt crops and refuge plants.[91]

It is difficult to understand the point of all this. Farmers who refused to comply with the government "recommendations" did not understand why they were expected to continue buying modified seed. Novartis dropped the price of the seed

[84]M. Mellon & J. Rissler, *Gone to Seed: Transgenic Contaminants in the Traditional Seed Supply*, Union of Concerned Scientists, Cambridge, MA (2004).

[85]E. C. Burkness, W. D. Hutchison, P. C. Bolin, D. W. Bartels, D. F. Warnock & D. W. Davis, 'Field efficacy of sweet corn hybrids expressing a Bacillus thuringiensis toxin for management of Ostrinia nubilalis (Lepidoptera: Crambidae) and Helicoverpa zea (Lepidoptera: Noctuidae)', *J. Econ. Entomol.* 94, 197–203 (2001).

[86]J. D. Sedlacek, S. R. Komaravalli, A. M. Hanley, B. D. Price & P. M. Davis, 'Life history attributes of Indian meal moth (Lepidoptera: Pyralidae) and Angoumois grain moth (Lepidoptera: Gelechiidae) reared on transgenic corn kernels', *J. Econ. Entomol.* 94, 586–592 (2001).

[87]B. E. Tabashnik & B. A. Croft, 'Managing pesticide resistance in *crop*-arthropod complexes: interactions between biological and operational factors', *Environ Entomol.* 11, 1137–1144 (1982).

[88]D. W. Onstad & C. A. Guse, 'Economic analysis of transgenic maize and nontransgenic refuges for managing European corn borer', *J. Econ. Entomol.* 92, 1256–1265 (1999).

[89]Y. Carrière, M. S. Sisterson B. E. Tabashnik, In: *Insect Pest Management: Field and Protected Crops*, R. Horowitz & I. Ishaaya (Eds), Springer, Berlin (2004), pp. 65–95.

[90]T. A. Horner, G. P. Dively & D. A. Herbert, 'Effects of MON810 Bt field corn on adult emergence of *Helicoverpa zea* (Lepidoptera: Noctuidae)', *J. Econ. Entomol.* 96(3), 925–930 (2003).

[91]C. F. Chilcutt & B. Tabashnik, 'Contamination of refuges by *Bacillus thuringiensis* toxin genes from transgenic maize', *Proc. Natl Acad. Sci. U. S. A.* 101(20), 7526–7529 (2004).

and launched promotions to ingratiate farmers and induce them to comply with the government programme.[92]

The hopes of the inventors of Bt-plants suffered another severe setback in the summer of 1996, when a heat wave hit the southern states of the US Monsanto Nu Cotn was growing vigorously and promising fabulous yields that would compensate the company for its many costs and the controversies in which it had been involved. As any farmer knows, heat and drought inhibit protein synthesis by plants, including the protein Bt. Moreover, the tobacco budworm proliferates under the same arid conditions. The two factors set the scene for disaster: the worm destroyed about half the crops in a few weeks.[93] In desperation, Monsanto advised farmers to spray what was left of the crops with conventional insecticides. Needless to say, the farmers were furious: the expensive Monsanto seeds were supposed to save them the cost and bother of synthetic pesticides.

And the trouble did not end there. The resistance acquired by insects exposed to Bt-plants, that express only 1 of 50 possible toxins, is not usually specific but unfortunately means cross-resistance to many types of endotoxins;[94] for example insects immune to CryIa(c) protein are also resistant to CryIA(a), CryIA(b), CryIB, CryIC(a) and CryIIA.[95] Since little is known about cross-resistance, which regards insecticides and also antibiotics, no preventive strategy is yet possible. Cross-resistance may appear after a single exposure to one of the many Bt-toxins[96] and cannot therefore be avoided by resorting to proteins belonging to the same family. Horizontal gene transfer presumably leads to incorporation of the sequence coding for Bt-toxin in the genome of plant progenitors of GM plants or even weeds, making them resistant to the insecticide and all the work that went into producing them in vain. Spread of Bt-resistance to weeds or unmodified ancestors increases selective pressure, making Bt completely ineffective in the span of a few years. This is not theory but a real threat, already observed in different cases, including those of potatoes and Brassicaceae.[97,98] Gene spread is a well-known and

---

[92] J. L. Fox, 'EPA seeks refuge from Bt resistance', *Nat. Biotechnol.* 15, 409 (1997).

[93] J. Kaiser, 'Pest overwhelm Bt cotton crop', *Science* 273, 423 (1996).

[94] Cross-resistance is when immunity to a drug or chemical compound automatically confers resistance to other drugs or molecules structurally similar to the original compound.

[95] B. E. Tabashnik, Y.-B. Liu, N. Finson, L. Masson & D. G. Heckel, 'One gene in diamond-back moth confers resistance to four *Bacillus thuringiensis* toxins', *Proc. Natl Acad. Sci. U. S. A.* 94, 1640 (1997).

[96] L. S. Bauer, 'Resistance: a threat to the insecticidal crystal proteins of *Bacillus thuringiensis, Florida Entomol.* 78, 414 (1995).

[97] M. D. Halfhill, R. J. Millwood, P. L. Raymer & C. L. Stewart, Bt-transgenic oilseed rape hybridization with its weedy relative *Brassica rapa'*, *Environ. Biosafety Res.* 1, 19–28 (2002).

[98] I. Skogsmyr, 'Gene dispersal from transgenic potatoes to con-specifics: a field trial', *Theor. Appl. Genet.* 88, 770–774 (1994).

inevitable phenomenon,[99] that is likely to augment with GMOs. Contamination of wild populations with transgenes has already occurred,[100] though accompanied by controversy aimed at minimizing its importance.[101] It is already taken for granted that new hybrids will appear from spontaneous crossing of modified and non-modified plants,[102] once engineered seeds are marketed:

> Thus, the relevant issue in assessing the risk of spread of transgenes is not hybridization probabilities, but rather the probability of spread and persistence of transgenes into wild plant populations. Transgenes can spread and persist in wild populations either through the back-crossing of transient transgenic hybrids with wild-type plants or by the stabilization and subsequent increase in the frequency of a transgenic hybrid line. These two events – i.e. successful introgression of wild-type plants or invasion by transgenic hybrids – are respectively considered a threat to the genetic diversity of wild plants and the biodiversity of natural communities.[103]

---

[99] P. Kareiva, W. Morris & C. M. Jacobi, 'Studying and managing the risk of cross-fertilization between transgenic crops and wild relatives', *Mol. Ecol.* 3, 15–21 (1994).

[100] D. Quist & I. H. Chapela, 'Transgenic DNA introgressed into traditional maize landraces in Oaxaca, Mexico', *Nature* 414, 541–543 (2001).

[101] N. Kaplinsky, D. Braun, D. Lisch, A. Hay, S. Hake & M. Feeling, 'Maize transgene results in Mexico are artefacts', *Nature* 416, 601–606 (2002). Subsequent studies documented F_1 hybridization between GM and native maize in a few areas (M. Metz & J. Futterer, 'Suspect evidence of transgenic contamination', *Nature* 416, 600–601 (2002)) and it seemed that the cross had become a stable introduction of the transgene into the host genome. This process is distinct from simple hybridization and requires various steps before introduction of the transgene becomes stable and can be transmitted to the progeny (C. N. Stewart Jr, M. D. Halfhill & S. I. Warwick, 'Transgene introgression from genetically modified crops to their wild relatives', *Nat. Genet.* 4, 806–817 (2003)). According to Luis Herrera Estrella (director of the Mexico Plant Biotechnology Centre), the contamination was due to mixing of native seeds for sowing and transgenic maize for food, imported from the USA, in percentages ranging from 1% to 35% (average 10%) in the lots examined (J. Hodgson, 'Maize uncertainties create political fallout', *Nat. Biotechnol.* 20, 106–107 (2002)). When Metz and Kaplinsky criticized the methods used, Quist and Chapela repeated the experiment by a new method (the first was based on inverse PCR, the second on DNA hybridization) that confirmed the contamination and also insertion of the transgene in the plant genome (D. Quist & I. H. Chapela, 'Quist and Chapela reply', *Nature* 416, 602 (2002)). The two authors received the support of many other researchers (Suarez, Strohman, Billings, Worthy) who criticized the derogatory attitude of the journal and wrote a long list of accusations about the criticism levelled at the original paper. Not only did they confirm the validity of the approach used by Quist and Chapela, but they also cast doubt on the intellectual honesty of seven out of eight of the "inquisition" who showed clear conflict of interests because their university had received about $25,000,000 from a subsidiary of Novartis, the company producing the maize in question. The signatories of the letter invited *Nature* itself to resolve its conflicts of interests, as it had clearly been influenced by pressure from the biotech industry ('Conflicts around a study of Mexican crops', *Nature* 417, 897–898 (2002)).

[102] N. C. Ellstrand, 'When transgenes wander, should we worry?', *Plant Physiol.* 125, 1543–1545 (2001).

[103] C. Vacher, A. E. Weis, D. Hermann, T. Kossler, C. Young & M. E. Hochberg, 'Impact of ecological factors on the initial invasion of Bt transgenes into wild populations of birseed rape (*Brassica rapa*)', *Theor. Appl. Genet.* 109, 806–814 (2004).

Experimental data shows that these new hybrids are extremely fertile and capable of adapting to a range of environments,[104] indeed, so efficient that they overwhelm the ecosystem in which they establish. This happened in China.

Sometimes real events confirm or even precede science fiction. It was recently discovered that years after introduction of large-scale cultivation of GM cotton, the economic gains ensured to Chinese growers evaporated. The reason, reported by researchers at Cornell University at the 2006 conference of the *American Agricultural Economics Association*, was that new parasites were devastating Bt-cotton crops. Bt-cotton was toxic to the cotton bollworm (*Helicoverpa armigera*), a major parasite of cotton, and accounted for about 25% of world cotton production.

> After seven years, however, the populations of other parasites increased so much that farmers were forced to spray fields with much greater quantities of pesticides than before the introduction of Bt cotton. The study – the first examining the long-term economic impact of Bt cotton – showed that for three years after 2001, farmers who sowed the biotech cotton reduced their pesticide consumption by more than 70%, earning 36% more than farmers who continued with conventional cotton. However, since 2004 they have been forced to return to the quantities of pesticides used on conventional cotton and earned 8% less than their competitors because of the higher cost of Bt seeds. The problem was not the development of parasites resistant to Bt cotton as originally thought, but that of other parasites insensitive to Bt toxin, previously controlled by broad spectrum pesticides. According to researchers, the emergence of these "secondary" parasites could be a serious problem for all countries where Bt cotton has been grown on a large scale: China, India, Argentina and the USA, though in the latter country farmers growing Bt cotton are also bound *by contract* to cultivate a limited area of traditional cotton, to provide refuge for *H. armigera*, where conventional pesticides are used.[105]

Introduction of Bt-plants caused sudden selective pressure in favour of predators resistant to Bt. The benefits of the GM varieties soon disappeared, farmers were saddled with greater costs and ecosystem equilibrium was upset.

The spread of the Bt-gene into the environment carries other risks, first of all those related to contamination of soil with Bt-toxins which are dangerous for other insects, including pollinators, predators of insect pests, soil organisms and

---

[104] M. Pertl, T. P. Hauser, C. Damgaard & R. B. Jorgensen, 'Male fitness of oilseed rape (*Brassica napus*), weedy *B. rapa* and their F_1 hybrids when pollinating *B. rapa* seeds', *Heredity* 89, 212–218 (2002).
[105] 'Notiziario', *Le Scienze*, 26 July 2006.

animals. Bt-protein applied as biopesticide disappears quickly from soil,[106,107] but that released by roots of Bt-plants continues to significantly inhibit larval growth for up to 120 days after dispersal;[108] at 140 days the biopesticide can still be detected, though in variable percentages of the initial dose (1–35%).[109] This is because Bt-protein quickly associates with organic microparticles that prevent its break-down by soil bacteria.[110] A study showed that transgenic DNA can survive in the soil under such conditions for up to 2 years, without being transmitted to plants.[111] In another study, transfer of genes conferring antibiotic resistance was docu-mented between strains of *Acinetobacter* and the microbe *Ralstonia solanacearum*, co-infectants of tobacco plants. This phenomenon seems to occur when the receptor bacterium has sequences homologous to those of the donor plant.[112,113]

In addition, Bt-transfection seems to be a very ineffective business. Indeed, high levels of expression of the *cry1Ac* gene from *Bacillus thuringiensis* cannot be routinely achieved in transgenic plants despite modifications made in the gene to improve its expression. This has been attributed to the instability of the transcript in a few reports. The expression of the Cry1Ac endotoxin has detrimental effects on both the *in vitro* and *in vivo* growth and development of transgenic plants. "A number of experiments on developing transgenics in cotton with different versions of *cry1Ac* gene showed that the majority of the plants did not express any Cry1Ac protein. Although the reasons for this detrimental effect need to be analysed, the study shows that the current scenario for developing transgenics that express optimal levels of the *cry* gene is not routine, due to the detrimental effects of high expression of these genes".[114]

---

[106] D. Saxena & G. Stotzky, 'Insecticidal toxin from *Bacillus thuringiensis* is released from roots of transgenic Bt corn *in vitro* and *in situ*', *FEMS Microbiol. Ecol.* 33(1), 35–39 (2000).
[107] G. Stotsky, 'Insecticidal toxin in root exudates from Bt corn', *Nature* 402(6761), 480 (1999).
[108] S. R. Sims & J. L. Ream, 'Soil inactivation of the *Bacillus thuringiensis* subsp. *kurstaki* CryIIA', *J. Agric. Food Chem.* 45, 1502–1505 (1997).
[109] C. J. Palm, R. J. Seidler, D. L. Schaller & K. K. Donegan, 'Persistence in soil of transgenic plant produced *Bacillus thuringiensis* var. *kurstaki* δ-endotoxin', *Can. J. Microbiol.* 42, 1258–1262 (1996).
[110] G. Venkateswerlu & G. Stotzky, 'Binding of the prototoxin and toxin proteins of *Bacillus thuringiensis* subsp. *kurstaki* on clay minerals', *Curr. Microbiol.* 25, 225–230 (1992).
[111] F. Gebhard & K. Smalla, 'Transformation of *Acinetobacter* sp strain BD413 by transgenic sugar beet DNA', *Appl. Environ. Microbiol.* 64, 1550–1554 (1998).
[112] E. Kay, T. M. Vogel, F. Bertolla, R. Nalin & P. Simonet, '*In situ* transfer of antibiotic resistance genes from transgenic (transplastomic) tobacco plants to bacteria', *Appl. Environ. Microbiol.* 68, 3345–3351 (2002).
[113] E. Kay, G. Chabrillat, T. M. Vogel & P. Simonet, 'Intergeneric transfer of chromosomal and conjugative plasmid genes between *Ralstonia solanacearum* and sp BD413', *Mol. Plant Microb. Interact.* 16, 74–82 (2003).
[114] P. Rawat, A. K. Singh & K. Ray, 'Detrimental effect of expression of Bt endotoxin Cry1Ac on *in vitro* regeneration, *in vivo* growth and development of tobacco and cotton transgenics', *J. Biosci.* 36, 363–376 (2011).

The process of release of Bt-protein by the roots of modified plants[115] can alter the soil ecosystem, though it is difficult to assess the impact of these interactions in time.[116] Some could impair the rate of plant decomposition and/or soil fertility, affecting nitrogen turnover.[117] In parallel, all this may involve a reduction in the biodiversity of soil microorganisms, which as we observed, may in turn impair soil yield.[118] The build-up of pesticide produced by the GM plants may cause bioaccumulation and affect the viability of other plants, insects and predators. The results obtained by the few studies conducted to date are contradictory:[119,120] the risk is confirmed and the need for further research is underlined. Despite many appeals, this research has not yet been undertaken.

Secondly, the gene coding for Bt also expresses other proteins, the action of which is unclear, but which are certainly toxic for some other species of insect.[121] This was announced emphatically by Novartis without realizing the concern it aroused in scientists familiar with the problem. If the modified gene synthesizes other proteins (at least three) besides Bt, about which nothing is known except that they are toxic, who can reassure us that they are not toxic for useful insects, other plants, animals or humans?

Rabbits in Australia

At the beginning of last century, an Old World couple moving to Australia had the idea of taking a pair of fertile young rabbits, along with their trunk of nostalgic sou-venirs. There were no rabbits in Australia and the young couple thought they would colonize the great southern land with a new species. No wish was ever granted so fully and so literally. The rabbits proliferated. The new environment posed no limitations: food was abundant, there were no predators and the newcomers soon numbered hundreds. Fatally, some escaped their pens and spread in all directions,

[115]D. Saxena & G. Stotzky, 'Bt toxin uptake from soil by plants', Nat. Biotechnol. 19(3), 199 (2001).
[116]B. S. Griffiths, L. H. Heckmann, S. Caul, J. Thompson, C. Scrimgeour & P. H. Krogh, 'Varietal effects of eight paired lines of transgenic Bt maize and near-isogenic non-Bt maize on soil microbial and nematode community structure',. Plant Biotechnol. J. 5(1), 60–68 (2007).
[117]M. Wood, Soil Biology, Chapman & Hall, New York (1989).
[118]M. G. A. van der Heijden, J. N. Klironomos, M. Ursic, P. Moutoglis, R. Streitwolf-Engel, T. Boller, A. Wiemken & I. R. Sanders, 'Mycorrhizal fungal diversity determines plant biodiversity, ecosystem variability and productivity', Nature 396, 69–71 (1998).
[119]H. A. Bell, 'Transgenic GNA expressing potato plants augment the beneficial bio-control of Lacanobia oleracea (Lepidoptera: Noctuidae) by the parasitoid Eulophus pennicornis (Hymenoptera: Eulophidae)', Transgenic Res. 10(1), 35–42 (2001).
[120]C. Zwahlen, A. Hilbeck, R. Howald & W. Nentwig, 'Effects of transgenic Bt corn litter on the earthworm Lumbricus terrestris', Mol. Ecol. 12(4), 1077–1086 (2003).
[121]Ciba-Geigy, Application for placing on the market a genetically modified plant (maize protecting itself against corn borers), B13, C 1.3.1, appendix C-8 (1994).

continuing to reproduce without a pause. In 10 years, the pair of rabbits had created a colony of hundreds of thousands of animals, soon to reach millions. Farmers and then the government began to grow worried. The rabbits destroyed everything in their path and became a pest for crops. In the space of months, land carefully culti-vated became wasteland. The reaction was to hunt them. Bands of hunters formed, miles of traps were set, tons of poison laid, all in vain. The rabbits eluded the traps and although thousands were shot or poisoned, they continued reproducing. Aus-tralia has by now been colonized, and the bosses are now those tender, harmless animals, symbol of cowardice, but famous for reproductive capacity. The battle was lost, until someone had the idea of spreading the *myxomatosis* virus on plants. This virus does not affect humans, but is lethal for various animals. There do not seem to have been any objections to this solution. The virus was disseminated and within weeks, the rabbits began to die like flies and their carcasses were trucked to special disposal sites. The invincible armada was soon defeated. The rabbit population was now reduced to more manageable proportions. The equilibrium upset by the introduction of a new species was restored by introducing a natural predator, which prevented the rabbits from invading the ecosystem. This example is instructive from various points of view.

First of all it reminds us that an equilibrium upset is followed by a cascade of reactions, many unpredictable. Secondly, it reminds us that introduction of a factor extraneous to the equilibrium can cause damage that if not counteracted by internal regulators can lead the whole system to catastrophe. In the case of rabbits in Australia, the existence of a natural predator without side-effects was a great fortune. Otherwise Australia would be even more of a desert than it is possible to imagine. Fortunately, the strengths and weaknesses of rabbits as a species were known. If instead of rabbits, Australia had been invaded by an unknown plant variety without any similar in nature, for which there were no chemical antidotes or natural predators, what would have happened? The plant could have supplanted all crops and other plants, making dozens of species extinct, irreversibly damaging the ecosystems whose equilibrium depended on those species: other plants would have disappeared and insects, worms, flowers and animals with them. Soil structure would have suffered: whole areas could not have been cultivated and the desert would probably have won. This is not a certainty, but it is a possible scenario. It suffices to illustrate the risks we run today in releasing modified species which represent a dangerous new item in the long list of contaminants of the modern world: genetic contamination.

Actually, comparison of uncontrolled spread of GMOs with that of synthetic chemical compounds is a simplification useful to illustrate a point. We have some understanding of the type of interactions chemical contaminants can cause and we know what to abide by as monitoring parameters are sufficiently standardized and precise to provide reliable estimates and predictions. However, in the case of genetic contamination, things are neither simple nor clear: GMOs are forms of life and their possible interactions with other lives are almost unlimited and certainly unpredictable. Like anything unknown, the risk is proportionally greater. GM structures have another worrying character: they move, they reproduce, they transmit modified genes to progeny and they have the bad habit of overcoming their natural relatives to become the dominant variant. In other words, bad money drives out good.

A similar scenario, though not due to transgenic food, has already been described in North American habitats by US researchers:

> Over the past several hundred years, thousands of non-native organisms have been brought to America from other regions of the world. While many of these creatures have adapted to the North American ecosystems without severe dislocations, a small percentage of them have run wild, wreaking havoc on the flora and fauna of the continent. Gypsy moth, Kudzu vine, Dutch elm disease, chestnut blight, starlings and Mediterranean fruit flies come easily to mind. For instance, the mongoose, introduced into Hawaii from India to control rodents who were damaging the sugar cane crop, became an environmental nightmare, devouring a wide range of native animals and, in the process, destabilizing the ecosystem of the islands. Also, the zebra mussel, a native of Europe, migrated to North America by attaching itself to ships and has become a formidable pest in the Great Lakes, blocking up water pipes at filtration plants and edging out native species in the Great Lake region.[122]

GM plants may compete with other varieties, becoming predominant and invasive. Introduction of foreign species may lead to invasions that cause structural and functional damage to natural ecosystems.[123] Invasive species cause damage estimated at 137 billion dollars per year in the USA and are considered among the top three environmental problems, together with loss of ecosystems and climate change.[124]

An undeniable fact is reported by scientists actively engaged in studying transgenesis:

> Our capacity to predict ecological impacts of introduced species, including GEOs, is imprecise, and data used for assessing potential ecological impacts have limitations. Our inability to accurately predict ecological consequences, especially long-term, higher-order interactions, increases the uncertainty associated with a risk assessment and may require modifications in our risk management strategies. [...] Additional or unidentified benefits and risks may exist that published data do not yet

---

[122] J. Rifkin, The Biotech Century, Penguin, Putnam, Inc., New York (1998).

[123] S. H. Reichard, 'Conflicting values and common goals: codes of conduct to reduce the threat of invasive species', WeedTechnol. 18, 1503–1507 (2004).

[124] D. Pimentel, 'Biological control of invading species', Science 289(5481), 869 (2000).

> address [...] Evaluation of potential environmental benefits [is]
> still in its infancy....[125]

The need for a risk assessment model that can provide transparent, unequivocal and coherent analysis on which to base economic and political decisions has been repeatedly expressed.[126] However, most studies so far conducted have been empirical[127] or qualitative.[128] The question remains unanswered: where is the proof and the scientific evidence that release of GMOs is risk-free?[129]

When GMOs are released into the environment, there is a real probability that the organism may become dangerous because it is an extraneous element, artificially introduced into an integrated network of relationships on which the ecosystem rests. The repercussions are unpredictable and cannot be corrected. Bernard Rollin, professor of physiology at the University of Colorado, writes:

> Experience teaches us that the dangers of releasing animals
> into a new environment cannot be estimated, even with species
> whose characteristics are well known. [...] Ignorance of what
> could happen with newly engineered creatures is even more
> certain.[130]

Indeed, it is not always possible to undo the damage. With the rabbits in Australia, it was possible to exploit their susceptibility to *myxomatosis*, but it is unlikely that similar solutions can be found for GM plants, animals or microorganisms appearing for the first time on the stage of nature.

A Little Ecology

The lack of clear standards of risk assessment is particularly evident in field test protocols that ought to set limits. These are experiments conducted with GMOs in plots of land with the aim of obtaining a precise assessment of ecological risks before the product is marketed on a large scale. Hailed initially as the scientific

---

[125]L. L. Wolfenbarger & R. R. Phifer, 'The ecological risks and benefits of genetically engineered plants', *Science* 290, 2088–2093 (2000).

[126]J. Beringer, 'Reply from Sir John Beringer', *Bull. Br. Ecol. Soc.* 31, 16 (2000).

[127]K. Pascher & G. Gollrnan, 'Ecological risk assessment of transgenic plants releases: an Australian perspective', *Biodiversity Conserv.* 8, 1139–1158 (1999).

[128]P. C. St Amand, D. Z. Skinner & R. N. Peaden, 'Risk of alfalfa transgene dissemination and scale', *Theor. Appl. Genet.* 101, 107–114 (2000).

[129]C. J. Thompson, B. J. P. Thompson, P. K. Ades, R. Cousens, P. Garnier-Gere, K. Landman, E. Newbigin & M. A. Burgman, 'Model-based analysis of the likelihood of gene introgression from genetically modified crops into wild relatives', *Ecol. Model* 162, 199–209 (2003).

[130]B. Rollin, *The Frankenstein Syndrome: Ethical and Social Issues in the Genetic Engineering of Animals*, Cambridge University Press, New York (1995), p. 119.

answer to environmentalists' criticisms, today the utility of field tests for their purported aim has been widely downsized.[131] Careful analysis of the limits of field tests has shown that the field simulation parameters were devised under artificial conditions calculated to make escape of pollen and seeds impossible or unlikely, thus the major risk associated with commercial production – escape of transgenic pollen through seeds – is excluded *a priori* in tests conducted on a limited scale.[132] The plots of land used for testing are small and "treated" to eliminate plants. One or two crops per season are cultivated. Clearly these temporal and spatial constraints do not permit resistance to arise, nor do they offer the concrete possibility of spread of modified genes into the environment. Such effects can probably only be detected when the new organisms are cultivated on a large scale, involving various ecosystems, for an appropriate period of time. Once these conditions are met, they render totally useless field tests conceived as indicators of environmental impact. Here lies one of the major paradoxes to which the introduction of biotechnologies exposes us: on one hand, proponents of GMOs rightly sustain that unless field tests are done on hundreds of thousands of hectares of land over various consecutive growing seasons, the results will not be reliable. On the other hand, and here we have the vicious circle, the results of transgenic dissemination are in any case irreversible and if the effects are harmful, the negative environmental repercussions cannot be undone. Under these conditions, the field tests make no sense, because their aim is to detect damage before it becomes substantial and irreparable. Field tests are meant to be preventive; in this case, they appear to be scientifically legitimate, but in fact they are not.

Actually, this is not the first time that such has been observed in the history of biotechnologies. Rigorous scientific assessment of the socio-environmental risks, damage and impact of GMOs has been persistently neglected. It is painful to have to point out that those who are so vocal about limits to the freedom of research are the first to abuse it, even omitting to take account of scientific data, as denounced in an eloquent editorial of *The Lancet*:

> The issue of genetically modified foods has been badly mis-handled by everyone involved. Governments should never have allowed these products into the food chain without insisting on rigorous testing for effects on health. The companies should have paid greater attention to the possible risks to health and of the public's perception of this risk; they are now paying the price of this neglect. And scientists involved in research into the risks of genetically modified foods should have published the results in the scientific press, not through the popular media;

---

[131]L. Firbank, M. Lonsdale & G. Poppy, 'Reassessing the environmental risks of GM crops', *Nat. Biotechnol.* 23(12), 1475–1476 (2005).
[132]A. A. Snow & P. M. Palma, 'Commercialization of transgenic plants: potential ecological risks', *Bioscience* 2, 95–103 (1997).

their colleagues, meanwhile, should also have avoided passing
judgments on the issue without the full facts before them.[133]

Unfortunately, the scientist's occupation has changed radically, and today the
key to fame and wealth implies other, often unmentionable, paths. A perverse network linking researchers, corporations and political-administrative control agencies
is the reason why the whole process from production to marketing of new GMOs
evades indispensable regulations that would considerably delay commercialization.
Any delay is to the detriment of investments, profits and the prestige accruing to
individuals and institutions. The objective of wealth at any cost does not provide
good counsel and portends unpleasant surprises, especially when decked in the
language of good intentions and inviolable principles. Thus it happened that while
a well-orchestrated media campaign was aimed at convincing and reassuring the
general public about the safety of GMOs, insurance companies (which see into the
future!) refused cover against risks associated with release of the new products,
well aware that no premium can ever cover the costs of an environmental disaster.
This is dramatic and undeniable evidence of the gravity of the side-effects to which
technologies based on manipulation of DNA expose us. Unlike many "scientists",
insurance brokers have a clear perception of the Kafkian dimensions implied by
attempts to regulate a technology without basic scientific knowledge about how the
new products could interact with the environment.

It is not necessary to establish absolutely whether GMOs are good or bad. It
would already be something if we had some elements on which to base judgement.
What we do know is that never in the history of humanity have there been experiments on such a large scale simultaneously involving so many parameters on which
ecological equilibrium, as we know it, is based.

For example, it is likely that at least 10% of new transgenic plants sown in
fields all over the world will behave like noxious weeds, invasive varieties that
displace plant species from their natural habitat, diminishing genetic variability of
ecosystems and upsetting their equilibrium. The inserted genes give the new plants
a competitive advantage in the short term. Resistance to pesticides, viruses, parasites and so forth threaten to make GM crops invasive through unusual advantages
with respect to their non-modified siblings. Plants resistant to cold could migrate to
other latitudes displacing original plants; early germinating plants could upset natural germination rhythms based on seasonal rotation of different plant varieties or
successfully colonize virgin land. A real example is rice modified to resist salinity,
a plant attracting great interest as a staple and offering producers the prospect of
conquering the vast markets of eastern Asia. What would happen if this type of rice
spreads along all the coasts? Indigenous populations would be displaced and other

---

[133]'Health risks of genetically modified foods', Editorial, *The Lancet* 353, 1811 (1999).

associated organisms such as algae, microorganisms, insects, arthropods, amphibians and birds may not be compatible with the invading rice. Different organisms, new to saltwater wetlands, could find homes in the new rice-dominated system.[134]

The main human food crops were domesticated over thousands of years, whereas industrial genetic modifications often break the natural barrier preventing gene transfer between different species and different kingdoms (between plants and animals) for the first time, increasing the probability of awakening wild characters such as those linked to invasiveness. Although proponents of GMOs deny the degree of risk, playing down the probability of events and repercussions,[135] it is well known that many apparently "mild" plants, such as barley, alfalfa, potatoes, wheat, sorghum and broccoli, have conserved their ancestors' capacity to become invasive, and this eventuality is the greatest threat associated with genetic engineering.[136]

As we have seen, the risk of transfer of genes conferring evolutionary advantages from modified plants to invasive varieties by cross-pollination is far from improbable. The problem is not new: in the 1980s, botanists and biologists feared that artificial hybrids would cause accentuated, uncontrollable gene flow. This is not probable but certain. For example, during experiments conducted under the control of the Danish Department of Science and Environmental Technology, GM herbicide-resistant canola was deliberately sown near a weed of the same species, *Brassica campestris*. As expected, cross-pollination occurred and the hybrid was fertile in 40% of cases, giving rise to invasive plants with herbicide resistance which will be difficult to extirpate.[137] The study showed that transgenic segments can easily spread from modified crops to nearby wild populations, causing genuine genetic contamination, the ecological, economic and legal implications of which are incalculable, as underlined in the *New York Times* article.

GMO corporations play down the danger by reassuring the public that the risk is low since transgenic crops normally do not grow near similar wild species and that gene flow only occurs over small distances. Unfortunately, the data produced by scientific research says something quite different. A study conducted over about one third of the Australian continent showed, for example, that spread of pollen from fields of GM canola travelled up to 3 km![138] Another control study on antibiotic-resistant potatoes assessed the possibility of contamination of normal potatoes

[134]J. Rissler & M. Mellon, *op. cit.*, p. 41.
[135]J. L. Azevedo & W. L. Araujo, 'Genetically modified crops: environmental and human health concerns', *Mutat. Res.* 544, 223–233 (2003).
[136]O. E. Sala, 'Global biodiversity scenarios for the year 2100', *Science* 287, 1770–1774 (2000).
[137]W. E. Learny, 'Gene inserted in crop plant is shown to spread to wild, *New York Times*, 7 March 1996, p. b14.
[138]M. A. Rieger, M. Lamond, C. Preston, S. B. Powles & R. T. Roush, 'Pollen-mediated movement of herbicide resistance between commercial canola fields', *Science* 296, 2386–2388 (2002).

growing at different distances from the experimental field. The modified gene was found in common potatoes in distant fields. One kilometre from the study field, 35% of seeds obtained from normal tubers contained the modified gene. The same is true for other crops such as pumpkin, turnips, sunflower, carrots and Bt-rice that can be spread in all directions by wind, at the risk of contaminating different wild species. This possibility has raised the concern of biologists and entomologists, the latter alarmed that super-species of resistant insects could emerge.[139]

The ease with which genes flow between and within species, combined with rapid globalization of trade and the explosion of international travel, make transport of contaminated weeds highly likely. Weeds infected by a wide range of genes have thousands of unthinkable dispersal routes that could involve the whole world in a unwanted experiment.

> Transnational life-science companies project that within 10 to 15 years, all of the major crops grown in the world will be genetically engineered to include herbicide-, pest-, virus-, bacterial-, fungus- and stress-resistant genes. Millions of acres of agricultural land and commercial forest will be transformed in the most daring experiment ever undertaken to remake the biological world. Proponents of the new science, armed with powerful gene-splicing tools and precious little data on potential impacts, are charging into this new world of agricultural biotechnology, giddy over the potential benefits and confident that the risks are minimum or non-existent. They may be right. But, what if they are wrong?[140]

Paradoxically, the plan to turn the world into a sort of biotech industry at the service of humans and the market, divorced from the context of laws that for millions of years have guided and established evolution, contains a basic contradiction that could lead to its failure.

Despite the daily press releases of the Biotechnology Trust, indiscriminately transmitted all over the world by mass media at the service of economic power, *genetic engineering cannot and will never be able to create a useful and original gene* ex novo *in the laboratory*. Biotechnology, like its precursors, is a greedy, predatory, extraction industry, no less than the mining industry, which is depleting the riches of the Earth and is incapable of producing anything itself. The primary materials come from nature and are extracted and possibly modified; they belong to the genome pool that the industry aims to appropriate to itself. This is the gold-rush of today, the race to grab the sequences on which the future of the biotech industry depends, by means of immoral and fraudulent patents.

---

[139] R. Weiss, 'Genetically engineered rice raises fear: as plants produce own insecticide, resistance buildup could occur', *The Washington Post*, 5 February 1996, p. A9.
[140] J. Rifkin, *op. cit.*

Sorry to say, the culprit is modern industry, first with the "green revolution" which has generalized the practice of monoculture and deforested vast and extraordinary areas of the planet, now with biotechnologies, which threaten the foundations of that wealth by destroying diversity. The standardization of GM crops, the spread of resistance to pesticides, the uncontrolled diffusion of genes are factors that cause extreme selective pressure, creating genetic uniformity and coercing biodiversity, a consolidated process in animal husbandry. According to Paul Raeburn, scientists can manipulate molecules and cells but cannot create even the simplest form of life in the laboratory. Only germplasm can ensure the continuity of life into the future. No advance in basic research can compensate the loss of the genetic material on which farmers depend.[141]

Genetic variability, which ensures the richness of germplasm, is indispensable for attenuation and buffering of attacks on the environment, offering a response to all types of structural fluctuations of ecosystems by providing individuals equal to a particular challenge by virtue of the diversity of their genomic heritage. Reduction of biodiversity causes damage that has well-known human and economic costs. It is curious that this does not raise alarm in the Church, which has always seen biodiversity as a sign of the wealth of the Divine project. According to St Thomas, God wanted the Earth to be populated by many diverse creatures with different levels of intelligence and different forms. Diversity and inequality ensured the orderly work of systems as if they were a single whole. St Thomas was convinced that if all creatures were the same, they could not act for the benefit of the others.[142]

The mutual dependence and undeniable hierarchy of life both rely on biological variability, that we are doing our best to reduce to an all-time low, a low dangerously close to the threshold of no return, the threshold below which life itself is threatened.

Those who see the history of humanity as a series of successes marking the ineluctable march of progress should bear in mind that never has agriculture been the victim of such profound devastation as it is now. The spread of monocultures, while good for the market economy, planning, increased yield and mechanization, has impoverished the genomic profile of selected varieties to such an extent that they have found themselves exposed to all kinds of diseases. Only by recovery of wild lines or naturally resistant strains have many crops been able to survive. An example is the European vine, destroyed by *Peronospora* in 1908 and recovered through the Californian strain. Coffee plantations were destroyed by disease in 1870, wheat by rust in 1904, rice in 1943 in India and maize in the 1970s in America. In all cases the

---

[141] P. Raeburn, *The Last Harvest: The Genetic Gamble that Threatens to Destroy American Agriculture*, University of Nebraska Press, Lincoln, NE (1995), p. 235.

[142] C. J. Glacken, *Traces on the Rhodian Shore: Nature and Culture in Western Thoughts from Ancient Times to the End of the Eighteenth Century*, Berkeley and Los Angeles University of California Press (1967), p. 230.

disaster was blamed on the practice of replacing complex ecosystems with selected monocultures, vulnerable to unexpected biological attack or for which there is no reservoir of natural strains expressing resistance.

The Irish potato famine is paradigmatic. From the second half of the 18th century, potatoes were the staple food of the poor in most European countries. In 1845, a mysterious epidemic that practically destroyed the potato crop hit Ireland for 5 consecutive years. More than a million persons died of hunger; the number forced to emigrate has never been calculated. Irish potatoes, like those of Europe, were direct descendents of a carefully selected domestic quality, very susceptible to disease. That line was literally swept away and resistant strains had to be found in the potato's country of origin (the Andes and Mexico) in order to restock Ireland.[143] The current spread of a limited range of selected and GM lines, while all other varieties are destined to disappear, exposes us to the possibility of a new epidemic for which we do not have effective remedies. Once the progenitors have disappeared from cultivation, where are we going to find substitutes?

The Unique Wealth of Biodiversity

So far only four GM varieties have effectively been marketed: canola, cotton, soybean and maize. Together they account for about 98% of the area cultivated with GM crops.[144] In the near future, however, second- and third-generation GMOs obtained by insertion of more than one transgene, and therefore with multiple characteristics, are likely to appear on the market:

> Predicting detrimental impact becomes more challenging as the diversity of GM releases grows and will be particularly difficult for transgenes that fundamentally change plant physiology (e.g. lignin content and drought tolerance). However, it is important to distinguish between unwanted environmental changes attributable to a transgene and those caused by other aspects of a dynamic agro-environment. [...] Current data on gene flow and spread are either insufficiently integrated or of an inappropriate scale to predict the likelihood or extent of transgene movement in a geographic region except in the broadest of terms.[145]

While some risks, including gene transfer, have been identified as likely, others can only be identified and quantified using an appropriate experimental model (tiered strategy):[146] the exact definition of risk is of capital importance and requires

---

[143]R. Rhoades, 'Incredible potato', *National Geographic*, May 1982, p. 679.
[144]Available at: http://www.isaaa.org.
[145]M. J. Wilkinson, J. Sweet & G. M. Poppy, 'Risk assessment of GM plants: avoiding gridlock?', *Trends Plant Sci.* 8(5), 208–212 (2003).
[146]J. Romeis, M. Meissle & F. Bigle, 'Transgenic crops expressing *Bacillus thuringiensis* toxins and biological control', *Nat. Biotechnol.* 24, 63–71 (2006).

assimilation of mutually compatible databases of many scientific disciplines. If independent studies can be useful in identifying risk and exposure on a small scale, it is difficult to imagine how they can be integrated to trace profiles for larger geographical dimensions.[147] Coordinated public research is therefore needed to obtain predictive measures of risk before release of the modified varieties. It is difficult not to agree with this advice by fierce sustainers of transgenesis, which has nevertheless still not been accepted. The concerns are therefore the same as before: biodiversity is threatened by GMOs.[148]

The impact of transgenic plants and their natural counterparts on biodiversity were compared in large-scale trials with maize resistant to herbicides, sugar beet and canola in Great Britain. Biodiversity proved to be impoverished in fields sown with transgenic beet and canola, but was higher in fields sown with transgenic maize, largely due to the fact that "conventional" farmers used atrazine, a strong herbicide banned in Europe.[149] When atrazine was replaced with other herbicides, the alleged promotion of biodiversity of GMOs proved insignificant.[150,151]

Thus the fears are not merely theoretical, but a fact that induced the FAO to formulate a policy of active protection of so-called green gold. Improvement of crop yields by traditional methods enabled less land to be cultivated.[152] This was accompanied by forced urbanization and a radical remake of farming practice, which called for mechanization, pesticides, fertilizers and irrigation. These changes, which were the core of what went down in history as the "green revolution", had major costs and often severely depleted natural resources and made many plant and animal species extinct.[153]

It is estimated that 40,000 plant species will become extinct by the middle of the present century. Thousands of others have already disappeared. The rate is 27,000 per year, 74 per day. In coastal Peru, dozens of wild varieties of tomato

---

[147] *Ibid.* Actually, early-tier tests were criticized because they did not consider chronic exposure or specific conditions that only occur in the field (A. Lang, E. La Uber & B. Darvas, 'Early-tier tests insufficient for GMO risk assessment', *Nat. Biotechnol.* 25, 35–36 (2007)).

[148] D. A. Andow, 'UK farm-scale evaluations of transgenic herbicide-tolerant crops', *Nature Biotechnol.* 21, 1453–1454 (2003).

[149] Atrazine is a chlorotriazine pesticide of the triazine group. Its use is banned in some European countries including Italy (since 1992), but in the world it continues to be one of the most widely used herbicides, selling 38 million kilograms per year.

[150] J. N. Perry, 'Ban on triazine herbicides likely to reduce but not to negate relative benefits of GMHT maize cropping', *Nature* 428, 313–316 (2004).

[151] S. A. Weaver & M. C. Morris, 'Risks associated with genetic modification: an annotated bibliography of peer reviewed natural science publications', *J. Agric. Environ. Ethics* 18, 157–189 (2005).

[152] G. Conway & G. Toenniesen, 'Feeding the world in the twenty-first century', *Nature* 402, C55–C58 (1999).

[153] J. R. Krebs, J. D. Wilson, R. B. Bradbury & G. M. Sirwardena, 'The second Silent Spring?', *Nature* 400, 611–612 (1999).

have already been lost.[154] In the Amazon, deforestation by oil companies has caused total loss of the various varieties of cocoa, among other things, of that immense region.[155] As rainforest destruction proceeds towards 50%, forecast for 2022, total extinction will regard many plants, estimated between 10% and 20% of those currently existing. The USA, generally reluctant to plunder wealth which does not belong to others, is losing about 5,000 of its original heritage of 25,000 plant varieties:[156] these include wild varieties of tomato, maize, peanuts, beans, peppers, apple melon and cocoa. For an idea of the rate at which this accelerated depletion is occurring, it was calculated that species became extinct at the rate of about one every 1,000 years at the time of the dinosaurs; by the industrial age, one species disappeared every 10 years; today three become extinct every hour.[157] This incalculable damage is objectively a threat to food security in the world, according to Edouard Saouma,[158] past director of the FAO. According to the capitalist logic of multinational exploiters of nature, anything that does not procure gain or profit, anything that cannot be "cultivated", is superfluous, useless, or even dangerous: dangerous because it prevents standardization of the land and uniformity of production to meet corporate needs, not dangerous for agriculture. The design of agrochemical corporations is for farmers to convert from mixed cultivation to monoculture and from there to growing exclusively GM crops. The result is that 97% of the crop varieties grown in the USA are now extinct: 86% of the more than 7,000 varieties of apples inventoried in 1905, more than 87% of 1,900 families of pear species. In India the situation is tragic: of the 30,000 varieties of rice once cultivated, today 10 species account for 75% of the harvest. Introduction of GM seeds has weakened genetic diversity and will continue to do so, by fuelling misplaced faith in too few plant genomes. The whole edifice of world agriculture, on which the survival of billions of persons depends, could collapse. The present course of action is equivalent to taking the foundation stone to repair the roof.[159] This is the opinion of many renowned biologists. Too bad the new engineers of life and their greedy bosses do not realize it.

I conclude this brief review with the words of one of the most militant, albeit honest and valid, proponents of GMOs, Renato Dulbecco. Dulbecco was awarded the Nobel Prize for his studies on cell cultures, and he is a strong promoter of

---

[154]C. Rick, Conservation of tomato species germplasm, *California Agriculture* (1977), p. 32 e ssg.

[155]J. Soria, Recent cocoa collecting expeditions, in: *Crop Genetic Resources for Today and Tomorrow*, O. H. Frankel & J. G. Hawkes (Eds), Cambridge University Press, New York (1975), p. 175.

[156]P. Raeburn, *op. cit.*, p. 167.

[157]E. Ecklom, *Disappearing Species: The Social Challenge*, Worldwatch Paper 22, Worldwatch Institute, Washington, DC (June 1978), p. 6.

[158]Quoted in J. Rifkin, *op. cit.*, p. 179.

[159]N. Myers, *A Wealth of Wild Species: Storehouse for Human Welfare*, Westview Press, Boulder, CO (1983), p. 24.

genetic manipulation in the biomedical field. He does not seem to view the experiments that his colleagues conduct in the field of agriculture and animal husbandry with the same favour. Indeed, he writes:

> Biotechnologies applied to agriculture arouse serious misgivings because they interact with the environment in unclear ways. We haven't the faintest idea what could happen if we alter the equilibria of the biosphere, releasing microorganisms created in the laboratory, that previously did not exist. Nothing may seem to happen for months or years, and then we may be confronted with something irretrievable. The gamble has too many unknowns. Only recently have we learned that nature cannot be violated with impunity and that disturbed equilibria re-establish to our detriment, in the form of floods, catastrophes, pollution or radioactivity. Ecological variations respond to undeniable needs, recognized by everyone. Even scientists must take them into account. We cannot act lightly when environmental equilibrium is at stake. We have to calculate the consequences of our actions and when that is not possible, when risks are unpredictable, forgo the experiment.[160]

These words are a rare example of intellectual honesty. But have the sustainers of GMOs read them?

---

[160] R. Dulbecco, *op. cit.*, pp. 24–25.

4

The chimera factory[1]

Nature will return, tho' you drive her out with pitchforks ...

Horace

Every day we sink a step deeper into putrid obscurity.

C. Baudelaire[2]

A Modern Bestiary

Texts known as bestiaries flourished in the Middle Ages. They were compendia of beasts, accompanied by a moral lesson. They included imaginary animals, such as chimeras, impossible collages of animal parts drawn by a demonic hand. The significance of these allegorical or esoteric figures could be interpreted on different levels. Bestiaries were often compiled by monks or scholars of arcane sciences. Behind an innocent façade, they sought to develop occult arguments and practices.[3] If someone, like my friend Roberto Marchesini,[4] had the idea of publishing a modern bestiary, he would not be taken for an outdated alchemist; his work would have the advantage of scientific veracity, since chimeras – monsters created by bioengineering – are already among us. Let us meet a few.

A Cat Called Birillo

As I write, my cat, a splendid British grey called Prince Guglielmo ("William"), *Birillo* for friends, walks on my desk and sorts my papers, playing with a pen, contemplating the computer screen and finally curling up for a nap. Earlier, he helped me eat a biscuit dunked in tea. *Birillo* is fond of my Japanese tea. I ask myself if we have ever really considered the sentiments, "thoughts" and emotions

---

[1] The title is borrowed from the book by R. Marchesini, *La Fabbrica delle Chimere*, Bollati Boringhieri, Torino (1999).
[2] C. Baudelaire, *Les Fleurs du Mal*: 'Chaque jour nous descendons d'un pas, à travers des tènebres qui puent.'
[3] F. Maspero, *Bestiario Antico*, Piemme, Casal Monferrato (AI) (1997).
[4] R. Marchesini, *op. cit.*

of our pets and of animals in general. Do we realize their importance in our lives, for the fact that they are simply what they are? Do they belong to us or do we belong to them? Could we ever intentionally harm them?

However, someone has plans for our silent and heroic friends, the many species of animals that have accompanied humans throughout history. Little attention has been paid to the suffering and degradation of animal species pursued cynically by the biotech industry for purposes that are not always noble and almost invariably inspired by aims that have nothing to do with knowledge or improvement of the human condition.

One of the first and most iniquitous consequences of the production of transgenic animals (i.e. animals in whose chromosomes one or more foreign genes have been inserted) is suffering. Insertion of a gene into a cell is not only completely random, but is forced into the nucleic acid matrix of the host by inoculation of hundreds of copies of the gene. This can cause anomalous insertions that lead to altered expression or deletion of regulatory segments, inhibiting or depressing protein synthesis, or directing production of abnormal, useless or deleterious polypeptides. This explains why hundreds or thousands of experiments are often necessary before a specimen capable of transmitting the transgene and the desired character to its progeny is successfully obtained. The transgene may even destroy natural genes of the host or cause unexpected mutations. One of the many examples described by Langley,[5] member of the Royal Society of Medicine of Great Britain and acute and untiring critic of biotech applied to animals, is the double insertion of a fruit fly gene and a gene from the Tk virus into mouse cells. Many mice born from this diabolical cross showed clear teratological damage: cleft muzzles, missing posterior limbs and brain aplasia. The aim of the experiments has never been declared, but the useless suffering inflicted on the animals was unmistakable.

Not only do transgenes give rise to unexpected effects but they often also have consequences that vary from organism to organism. For example, another study concerned mice transfected with a portion of the genome of virus SV40, an "inactivated" oncogenic virus, associated with a gene sequence that promotes cell replication in the chambers of the heart. The results were not completely negative but researchers were disappointed. They were surprised to find that the two halves of the heart grew at different rates: the right atrium grew a hundred times larger than the left and ended up killing the mice, invading their defenceless bodies. It was postulated that a third element must help regulate expression and function of the transgene, hitherto regarded as omnipotent. Atrial natriuretic factor, a biochemical messenger that exerts feedback on DNA, was suspected to have influenced gene

---

[5] G. Langley, A critical view of the use of genetically engineered animals in the laboratory, in: *Animal Genetic Engineering*, Wheale and McNally (Eds), London (1995), pp. 184–201.

expression in a different and selective way.[6] The episode illustrates the complexity of the interactions to which genes are subject: genes *per se* do not specify anything outside their original context. In context, they have a role and are coherent. The episode shows that insertion of a DNA segment not only fails to ensure any result, but exposes the organism to unexpected, often dramatic, consequences. Those who think that human diseases can be conquered by inserting the right gene, will find food for thought, and hopefully a change of opinion, in the following story.

Turn Off the TV and Stop Reading the Newspapers

Those accustomed to following the developments of research and the biotech industry in the media are in for a shock if they turn to specialist journals. The real situation is quite different from that described in daily newspapers and the glossy pages of popular science, where science is portrayed as a never-ending sequence of victories wrested from hostile Nature. The path of science is scattered with difficulties, errors, dashed hopes, stalemates, sudden changes of direction and lies advertised for various reasons.[7] The following story is an example.

At the end of the 1980s, it was proposed to modify pig embryo cells to induce them to produce industrial quantities of human growth hormone. The aim was to accelerate the growth of pigs, so that their weight doubled much faster. Since the 1970s this aim has been pursued systematically for many animals, from cows to chickens, but it was initially sought by supplementing animal diet with vitamins or low doses of antibiotics, or by intravenous administration of growth hormone. Scientists of the USDA in Maryland were bitterly disappointed by the results provided by the genetically modified pigs, which showed strange immune system and skeletal muscle anomalies.[8] Instead of a super-race of strong muscular pigs, researchers were confronted with a litter of pigs that were indeed giant, but weak and wispy, almost incapable of movement. They were condemned to a short and painful life. The research team had to admit that it was unable to control the gene coding for growth hormone in the way it had hoped.[9] The project was therefore varied, embarking on the idea of inserting a chicken, rather than a human gene in pig embryos. This was expected to produce pigs with robust muscles and "broad

[6]B. E. Zimmerman, 'Human gene line therapy: the case for its development and use', *J. Med. Phil.* 12, 594–598 (1991).

[7]F. Di Trocchio, *Le Bugie della Scienza*, Mondadori, Milan (1993).

[8]R. S. Cowan, Genetic technology and reproductive choice – an ethics for autonomy, in: *The Code of Codes: Scientific and Social Issues in the Human Genomic Project*, D. Kelves & L. Hood (Eds), Harvard University Press, Cambridge, MA (1995), p. 245.

[9]A. Lippman, 'Mother matters: a fresh look at prenatal genetic testing', *Issues Gen. Eng.* 2, 142 (1994).

shoulders": the famous Arnie Schwarzenegger pigs,[10] as Vern Pursel, the biologist responsible for the first experiments, called them.

We still do not know the outcome of this ambitious research,[11] but we know what happened to chickens modified with the growth gene of cows. As everyone knows, the chickens of yesterday, the rustic free-range birds of superlative flavour and unmistakable texture, are now only a memory. The chickens sold in supermarkets are swollen with water, flaccid and stuffed with antibiotics and hormones that inflate a chick to 2–3 kg in less than 7 weeks. With traditional methods, it took at least 6 months to reach this weight! The new transgenic chicken is no better than its hormonally doped predecessor. Planned to "develop" fast, with lean meat and early sperm production, the Chicken-chick is susceptible to malformation of the limbs, which cannot keep up with the growth of soft tissue, as well as to many diseases, development anomalies and reduced survival.[12]

The unhappy fate of the mouse, engineered by inserting the growth hormone gene, was not so different. Big fat mice were expected and the poor animals were indeed such; however in addition they had severe liver and kidney damage, structural abnormalities of the heart and spleen, an elevated incidence of tumours, extraordinarily high juvenile mortality, reduced fertility and in general a significantly shorter life expectancy than non-modified controls.[13]

Unfortunately, the monstrosities induced by transgenesis are seldom immediately evident. The first two generations of pigs with the GH gene did not show any apparent damage; subsequent generations had gastroduodenal ulcers, cardiomegaly, arthritis, nephritis and infertility.[14] In other species, the problems do not appear until the modified gene or genes are represented on both alleles (homozygosis). For example, mice modified with the gene coding for heat shock protein 70 (hsp70) from *Drosophila* or herpes virus, in order to synthesize thymidine kinase, are phenotypically normal in the first generation (heterozygotes), but the second

---

[10]M. A. Rothstein, Ethical issues surrounding the new technology as applied to health care, in: *Biotechnology*, Rundolph and McIntire (Eds), p. 204.

[11]Recently our friend Prof. Antonio Musarò successfully created a "Schwarzenegger mouse" for experiments on the impact of microgravity on muscles. The research does not have agricultural or economic aims but seeks to discover a solution to the muscle atrophy affecting astronauts (cf. A. Musarò, K. McCullagh, A. Paul, L. Houghton, G. Dobrowolny, M. Molinaro, E. R. Barton, H. L. Sweeney & N. Rosenthal, 'Localized Igf-1 transgene expression sustains hypertrophy and regeneration in senescent skeletal muscle', *Nat. Genet.* 27, 195–200 (2001)).

[12]P. Kitcher, *The Lives to Come: The Genetic Revolution and the Human Possibilities*, Simon and Schuster, New York (1996), p. 71 e ssg.

[13]E. Wolf & R. Wanke, Growth hormone overproduction in transgenic mice: phenotypic alterations and deduced animal models, in: *Welfare Aspects of Transgenic Animals*, L. F. M. van Zutphen & M. van der Meer (Eds), Springer-Verlag, Berlin (1995), pp. 26–47.

[14]V. G. Pursel, C. A. Pinkert, K. F. Miller, D. J. Bolt, R. G. Campbell, R. D. Palmiter, R. L. Brinster & R. E. Hamer, 'Genetic engineering of livestock', *Science* 244, 1281–1288 (1989).

generation (homozygotes) shows atrophy of the gluteus muscles, deformed limbs, cleft palate and olfactory lobe alterations.[15,16] Often the new genotype is far from stable even after many generations. A study of transgenic mice (lck-IL-4), modified to express high levels of the cytokine IL-4, showed severe osteoporosis with spinal gibbus deformation after various generations. The unexpected result was probably due to chromosomal rearrangement of the gene insert to unpredicted new sites.[17]

It is therefore certain that animals suffer in a new and unjustified way. As far as the quality and safety of the meat that reaches our tables is concerned, everyone is free to form his own opinion. Here too, some aspect of gene control is not working, and we are unlikely to discover what. The many doubts and fears include the possibility that transgenesis may facilitate production of anomalous proteins that trigger new diseases caused by prions.[18] It will probably not be possible to develop techniques that can check all factors involved in the regulation of functional expression of genomes within a reasonable time. Who can therefore reassure us that events of this type do not happen again? Cloning is a process loaded with unknowns and dangers: perhaps it is not a coincidence that the alarm about prions[19] was launched by the person who opened Pandora's box by making the first clone: an unfortunate sheep named Dolly.[20]

Milk with Hormones

Problems for animal and human health do not depend solely on insertion of modified genes, but also on relatively carefree use of the products of those genes. Today genetic engineering enables cheap industrial production of a quantity of hormones that once would have had to be extracted and purified by complex, expensive and relatively inefficient techniques. This is certainly an important advance that has provided doctors with a whole new arsenal of drugs for therapies that were once unimaginable. However, these products are not always used in commendable ways. A good example is bovine growth hormone, or bovine somatotropin (Bst), produced by a bacterium, *Escherichia coli*, in whose genome the bovine gene coding for the

---

[15] E. D. Murphy & J. B. Roths, 'Purkinje cell degeneration, a late effect of beige mutations in mice', *Annu. Rep. Jackson Lab* 49, 108–109 (1978).

[16] J. D. McNeish, W. J. Scott & S. S. Potter, 'Legless, a novel mutation found in PHTI – 1 transgenic mice', *Science* 241, 837–839 (1988).

[17] D. B. Lewis, H. D. Liggitt, E. L. Eftmann, S. T. Motley, S. L. Teitelbaum, K. J. Jepsen, S. A. Goldstein, J. Bonadio, J. Carpenter & R. M. Perimutter, 'Osteoporosis induced in mice by overproduction of interleukin 4', *Proc. Natl Acad. Sci. U. S. A.* 90, 11618–11622 (1993).

[18] P. R. Wills, 'Transgenic animals and prion diseases', *New Zealand Vet. J.* 43, 86–87 (1995).

[19] R. Jaenisch & I. Wilmut, 'Don't clone humans', *Science* 291, 2552 (2001).

[20] I. Wilmut, E. Schnieke, J. McWhir, A. J. Kind & K. H. Campbell, 'Viable offspring derived from fetal and adult mammalian cells', *Nature* 385, 810–813 (1997).

hormone has been inserted. Bst is principally used in animal husbandry, where it is administered to cows to increase milk production. This aim is questionable, considering that in Europe, milk production greatly exceeds market demand and the European Community is forced to pay incentives for destruction of surplus each year, in an attempt to sustain producers and maintain milk prices. In the USA, the situation is similar, with the government spending more than a billion dollars per year to absorb surplus milk production. It is thought that use of Bst will increase production by 12%. A degree in economics is not necessary to realize that this trend, decreasing demand (the human birth rate is declining) and unchanged public intervention, will soon lead to a drop in prices. Increasing production with falling prices is a certain recipe for bankruptcy of small producers and concentration of the milk sector in few greedy hands: the same hands that planned the industrial production of animals at the expense of the environment and product quality. Meat, eggs, milk and its derivatives are produced like any other industrial good, endeavouring to improve the appearance and competitivity on the basis of parameters and criteria that have nothing to do with nutritional or culinary needs and by techniques whose impact on the health of humans and animals has never been seriously considered.

If food has become a good, animals are turning into formless machinery without souls or any intrinsic value, appreciated only for their capacity to produce goods that can be manipulated. Since the second world war, productivity has been enhanced by forcing animals to sustain rates and living conditions worthy of the worst concentration camp of Stalin, and to increase meat and milk yields by illicit means: natural feed gave way to feedstocks based on animal flour (forgetting that bovines are ruminants that can only convert plant proteins), spiked with low doses of antibiotics (which for unknown reasons promote weight increase), vitamins and above all beta-blockers and hormones. The endocrine factors involved are of different chemical types, but all belong to the class of steroids, the same that in humans are responsible for various diseases, including cancer.

Europe long opposed the marketing of beef with hormones and the import of meat of animals treated with endocrine drugs, mostly from USA, Australia and Canada. This was appropriate but was not backed by true conviction nor was it accompanied by any statement of principle. Too bad, because any such statement would have been difficult to contradict.

It turned out that the USA challenged Europe at the WTO, accusing Europe of commercial protectionism of continental animal products, since according to the Americans, there was no scientific evidence demonstrating that US meat and the hormones used to produce it were dangerous. Of course, the WTO commission found in favour of the USA. Thus, while doping, bodybuilding and drug abuse among athletes and adolescents are condemned, the use of substances equally dangerous for health, to raise animals whose meat and milk will appear on our tables, is fostered and legitimized.

The hormone in question, Bst or rBgh, recombinant bovine growth hormone, was the first transgenic product marketed. It was released by Monsanto as Posilac and is currently used to increase milk production by about 20% in intensive animal farms in the USA, Brazil, Mexico, China, Egypt and India. For the animals, Bst acts as a drug, weakening muscles, altering metabolism, inducing prolonged energy imbalance[21,22] and impairing bone structure. Indeed, to ensure high milk production, the cow is forced to recover calcium and fluids from storage sites such as bone. Milk with hormones, in fact, has significantly lower concentrations of vitamins, proteins and minerals than milk of untreated cows.[23] Above all, overstimulation of the pituitary axis and udders causes a tremendous increase in ovarian cysts, infertility and mastitis. About 80% of cows are more susceptible to infection, leading to more frequent and massive treatment of the animals with antibiotics. The use of antibiotics in dairy cattle contributes to the spread of resistance to antibiotics by a broad spectrum of microbes,[24,25] and bacteria resistant to one or more antibiotics are often found in raw sausages and cheese.[26] Few people imagine that yoghurt and milk contain antibiotics![27,28]

According to the Pure Food Campaign, Bst produces sick cows that produce unhealthy milk and meat that could affect consumer health.[29] The leaflet accompanying phials of Bst specifies that use of Bst may cause an increase in the incidence of mastitis and of cows with mastitis. It also increases the rate of subclinical infections that are not immediately detected. In certain cases, use of Posilac is associated with

---

[21]R. H. Phipps, A review of the influence of somatotropin on health, reproduction and welfare in lactating dairy cows, in: *Use of Somatotropin in Livestock Production*, E. Sejrsen (Ed.), EEC/Elsevier (1989), pp. 88–119.

[22]M. Mason, 'Milk? It may not do a body good', *The Washington Post*, 7 March 1994.

[23]S. S. Epstein, 'Potential public health hazards of biosynthetic milk hormones', *Int. J. Health Serv.* 20(1), 73–77 (1990).

[24]V. Perreten, F. Schwarz, L. Cresta, M. Boeglin, G. Dasen & M. Teuber, 'Antibiotic resistance spread in food', *Nature* 389, 801–802 (1997).

[25]D. J. Chadwick & J. Goode (Eds), *Antibiotic Resistance. Origins, Evolution, Selection and Spread*, Ciba Foundation Symposium 207, Wiley, Chichester (1997).

[26]M. Yamaki, M. I. Berruga, R. L. Althaus, M. P. Molina & A. Molina, 'Screening of antibiotic residues in ewes' milk destined to cheese by a commercial microbiological inhibition assay', *Food Additives Contam.* 23(7), 660–667 (2006).

[27]M. Yamaki, M. I. Berruga, R. L. Althaus, M. P. Molina & A. Molina, 'Occurrence of antibiotic residues in milk from Manchega Ewe dairy farms', *J. Dairy Sci.* 87, 3132–3137 (2004).

[28]S. Bogialli, A. Di Corcia, A. Laganà, V. Mastrantoni & M. Sergi, 'A simple and rapid confirmatory assay for analyzing antibiotic residues of the macrolide class and lincomycin in bovine milk and yoghurt: hot water extraction followed by liquid chromatography/tandem mass spectrometry', *Rapid Comm. Mass Spectrom.* 21, 237–246 (2006).

[29]Pure Food Campaign, *What's Wrong with Genetically Engineered Foods?*, Washington, DC (1005), p. 134.

an increase in the number of somatic cells, or pus,[30,31] in milk. This possibility is admitted by the FDA but evidently does not worry the directors of the organization. The scientists charged with assessing the safety of Bst drew conclusions that did not meet the expectations of the producer and permission to publish the findings was denied:[32] this episode illustrates the meaning of "freedom of research" when the results conflict with vested interests!

The fact that cows must now be treated with increasing doses of antibiotics and other drugs for diseases related more or less directly to administration of Bst creates conditions for the development of strains of bacteria resistant to antibiotics and increases the likelihood of drug residues in milk and other human food. It is impossible to determine which and how many drugs are administered to animals: the FDA and corresponding European institutions only stipulate analytical procedures for four types of antibiotics, which amounts to consenting to illicit use of antibiotics and other compounds for which no analysis is necessary. It is therefore not surprising that in the course of inspections in the period 1990–1992, no less than 70 different unauthorized drugs were found on the premises of dairy companies: 40 of the drugs were not even authorized for veterinary use.[33] The US Government Accounting Office realized the risks associated with use of Bst for humans and animals and on 2 March 1993 reported to the Ministry of Health that the increased incidence of mastitis reported in research into Bst suggested that the possibility of an increase in antibiotics in milk was real. It added that foods containing Bst or derivatives of animals treated with Bst should not be approved until the risk had been correctly evaluated.[34] The FDA ignored the report, authorizing Monsanto to sell Posilac without requesting any specific documentation; the Department of Agriculture (USDA) avoided any interference, attempting to hide or minimize the results of scientific research.[35]

A final aspect of intensive use of Bst arouses concern among oncologists. Recombinant bovine growth hormone is not exactly identical to the original

---

[30] Cited in A. Apoteker, L'invasione del pesce-fragola, Editori Riuniti, Rome (2000), p. 54.

[31] T. C. White, K. S. Madsen, R. L. Hintz, R. H. Sorbet, R. J. Collier, D. L. Hard, G. F. Hartnell, W. A. Samuels, G. de Kerchove & F. Adriaens, 'Clinical mastitis in cows treated with sometribove (recombinant bovine somatotropin) and its relationship to milk yield', J. Dairy Sci. 77, 2249–2260 (1994).

[32] In a statistical review of the art commissioned by the producers of Bst, Brunner and Millstone found evidence that Bst promotes mastitis but could only publish a summary of their work, despite the fact that it was accepted by a peer-reviewed journal, because the producers opposed publication of the results and threatened legal action (see E. Brunner & E. Millstone, The Lancet 354, 71 (1999)).

[33] M. Hansen, Testimony before the Joint Meeting of the Food Advisory Committee on whether to label milk from rBgh-treated cows, New York University Press (1993).

[34] Cited in A. Apoteker, op. cit., p. 53.

[35] E. Millstone, E. Brunner & I. White, 'Plagiarism or protecting public health?', Nature 371, 647–648 (1994).

molecule, differing due to the presence of an extra amino acid, methionine, attached to the end of the polypeptide chain. This difference is not negligible because it can lead to different macromolecule folding, modifying its function and imparting new properties such as allergenicity. It may not be a coincidence that a rapidly increasing number of persons, not only children, are developing strange intolerances to milk proteins:[36,37] these intolerance now affect about 75% of consumers. Bst may also play a major role in the development of breast cancer, since it significantly increases secretion of insulin growth factor 1 (IGF-1) in cow's milk. Bovine IGF-1 is structurally identical to that of humans and has various known functions, such as differentiation and growth, but it is thought to participate in some ways in fostering degenerative diseases, like diabetes and breast cancer.[38,39] Studies to assess the safety of Bst and IGF-1 were conducted by the FDA, together with milk producers, but the results were never published. The abstract[40] indicates a number of methodological defects and above all the impossibility of evaluating long-term effects of consumption. Indeed, mice modified to express concentrations of IGF-1 20 times higher than normal have a high incidence of tumours after only 18 months.[41] IGF-1 is not destroyed by any normal procedure such as pasteurization, gastric digestion or boiling, and is present in concentrations 10 times higher in milk of cows treated with Bst than in milk of control cows. The risk of increased IGF-1 secretion has raised protest and alarm in the American Medical Association,[42] but again the federal institutions and milk producers successfully silenced the obscurantist Cassandras.

Not Only Cows Are Mad

The case of mad cow disease is paradigmatic in that it illustrates the dangers connected with two orders of themes related to modern animal husbandry, while not directly related to GMO. It shows the limits of carefree use of technologies

[36]L. P. Shek, L. Bardina, R. Castro, H. A. Sampson & K. Beyer, 'Humoral and cellular responses to cow milk proteins in patients with milk-induced IgE-mediated and non-IgE-mediated disorders', *Allergy* 60(7), 912–919 (2005).

[37]S.L. Bahna, 'Cow milk allergy vs cow milk intolerance', *Ann. All. Asthma Immunol.* 89(Suppl.), 56–60 (2002).

[38]C. Sell *et al.*, 'Simian virus 40 large tumor antigen is unable to transform mouse embryonic fibroblast blocking IGF-1 receptor', *Proc. Natl Acad. Sci. U. S. A.* 90, 11217 (1993).

[39]J. L. Outwater, A. Nicholson & N. Barnard, 'Dairy products and breast cancer: the IGF-I, estrogen and bGH hypothesis', *Med. Hypotheses* 48, 453–461 (1997).

[40]J. C. Juskevich & C. G. Cuyer, 'Bovine growth hormone: human food safety evaluation', *Science* 249, 875–884 (1990).

[41]C. E. Rogler, D. Yang & L. Rossetti, 'Altered body composition and increased frequency of diverse malignancies in insulin-like growth factor-II transgenic mice', *J. Biol. Chem.* 269(19), 13779–13784 (1994).

[42]American Medical Association Council of Scientific Affairs, 'Biotechnology and the American Agriculture Industry', *JAMA* 265, 1429 (1991).

based on wrong assumptions (such as "all information resides in DNA") and having unknown consequences; it also reveals the monstrous, anti-ecological nature of modern systems of raising animals.

Mad cow disease or BSE and many other degenerative diseases of the human and animal nervous system are due to the action of prions, infectious particles consisting solely of protein. Prions are not viruses as they do not have DNA or RNA. They cause inherited as well as transmissible diseases. A single protein can multiply in unpredictable ways, transforming normal protein molecules into dangerous ones, simply by a change in conformation.

Prion diseases, improperly known as "slow viral infections" because of the long latency between infection and the first symptoms, are particularly widespread in animals. The most common example is scrapie.[43] Years or decades after infection, the brain cortex of affected animals degenerates, developing holes and cavities. The animal loses coordination of movements and can no longer stand on its feet, accompanied by such intense itching that it scratches away part of its fleece (hence the name scrapie from the verb *to scrape*).

Other diseases caused by prions are transmissible mink encephalopathy, feline spongiform encephalopathy and BSE. Mad cow disease was first identified by G.A. Wells and J.W. Wilesmith in the late 1980s in England. The cause was pinpointed as addition of flour, made from sheep flesh and bones, to cattle feed.[44] The method used to prepare feedstocks until the 1970s eliminated the agent of scrapie, but after that date, process modifications to save time and money did not guarantee destruction of prions.

Human prion diseases are more complex and mysterious. For example, Kuru or laughing death, a form of brain degeneration with ataxia and dementia, first described in 1957, affected cannibals of Papua New Guinea. A certain tribe practised a funeral ritual in which the brains of the deceased were eaten. Since the practice was banned, Kuru has almost disappeared.[45]

Creutzfeldt–Jakob (CJ) disease, on the other hand, is ubiquitous and has a rate of one in a million persons over 60 years of age. It manifests as a particular type of dementia, usually following iatrogenic infection (after corneal transplant

---

[43] R. M. Anderson, C. A. Donnelly, N. M. Ferguson, M. E. Woolhouse, C. J. Watt, H. J. Udy, S. MaWhinney, S. P. Dunstan, T. R. Southwood, J. W. Wilesmith, J. B. Ryan, L. J. Hoinville, J. E. Hillerton, A. R. Austin & G. A. Wells, 'Transmission dynamics and epidemiology of BSE in British cattle', *Nature* 382(6594), 779–788 (1996).

[44] G. A. Wells & J. W. Wilesmith, 'Bovine spongiform encephalopathy', *Curr. Top. Microbiol. Immunol.* 172, 21–38 (1991).

[45] E. McKintosh, S. J. Tabrizi & J. Collinge, 'Prion diseases', *J. Neurovirol.* 9(2), 183–193 (2003).

or treatment with growth hormone from the human pituitary) or by inheritance. Other prion diseases affecting humans include fatal familial insomnia characterized by chronic insomnia, dementia and death, and Gerstmann–Straussler–Sheinker disease characterized by ataxia and dementia.[46]

These diseases are due to a protein normally synthesized in the body, PrP (prionic protein), first discovered in the brain by S.B. Prusiner in 1984.[47] PrP may have anomalous spatial conformation, known as beta filaments. Anomalous PrP can induce a similar change in normal molecules, quickly "infecting" the whole body. In some prion diseases (e.g. fatal familial insomnia), synthesis of anomalous PrP depends on alterations of the gene that synthesizes the normal protein. In other cases (such as BSE), PrP enters the body with food and many years later may trigger progressive transformation of normal protein. The mechanism can be likened to the bad apple that makes the whole basket of apples go bad.[48] To the humiliation of molecular biologists, this process occurs without involving nucleic acids (DNA or RNA).[49]

The first PrPs of animal origin discovered probably did not affect humans, since there is no experimental proof of transmission between species. Indeed, a barrier effect prevents proteins of distant animal species from inducing prion diseases. For example, if Syrian hamster PrP is inoculated in mice, the latter rarely develop the disease. However, if cows are fed with flour from sheep with scrapie, the disease is transmitted because sheep and cows have similar genomes. On the basis of this information, the probability of transmitting animal PrP to humans is low.

This belief enabled government authorities to sleep peacefully for many years. They were awoken suddenly by an unexpected variant of CJ disease identified in England in 1987. The prion isolated from humans showed surprising similarities to that extracted from brain specimens of cows with BSE.[50] From morning to night, it was realized that yet another barrier had been broken; after the many that humans had broken with impunity, the much feared possibility of a terrible epidemic was now real. This demonstrated that the behaviour of proteins and nucleic acids is so unpredictable that nobody can play with the bricks of life, relying on barriers

---

[46] Various authors, 'Le malattie da prioni', *Le Scienze* 393, 36–57 (2001).

[47] K. Kaneko, D. Peretz, K. M. Pan, T. C. Blochberger, H. Wille, R. Gabizon, O. H. Griffith, F. E. Cohen, M. A. Baldwin & S. B. Prusiner, 'Prion protein (PrP) synthetic peptides induce cellular PrP to acquire properties of the scrapie isoform', *Proc. Natl Acad. Sci. U. S. A.* 92(24), 11160–11164 (1995).

[48] B. Caughey, D. A. Kocisko, G. J. Raymond & P. T. Lansbury Jr, 'Aggregates of scrapie-associated prion protein induce the cell-free conversion of protease-sensitive prion protein to the protease-resistant state', *Chem. Biol.* 2(12), 807–817 (1995).

[49] P. T. Lansbury Jr & B. Caughey, 'The double life of the prion protein', *Curr. Biol.* 6(8), 914–916 (1996).

[50] S. B. Prusiner, 'Le malattie da Prioni', *Le Scienze* 319, 22–30 (1995).

between species. In other situations, every effort had been made to overcome these barriers! "Those who live by the sword shall perish by the sword."

Five years later, the incidence of the new variant of CJ is fortunately still low, about a hundred cases in England, three in France and only one in Ireland.[51,52] But how many cases should we expect? We know almost nothing about incubation times and this prevents us from making reliable forecasts, as well as keeping us in suspense. Are we still in time to prevent the irreparable?

The madness of cows, forced to become carnivores at the service of profit, was a herald of the madness that would affect those who ate the meat, and reflects the madness of completely "dehumanized" animal husbandry modelled on standardization of livestock:

> Breed will be replaced by the genetic model, much more convincing in ensuring morphofunctional and behavioural standards. In other words, cloning [and other genetic modifications] will completely transform animals into objects or machines. This has so far been prevented by population variability with its element of unpredictability.[53]

Needless to say, such a process is exactly the opposite of what happens in Nature. Nature rewards biodiversity. Biological evolution is based on biodiversity. The concept expressed by the forgotten "fathers" of molecular biology, first and foremost Francois Jacob,[54] for whom there was a close relationship between the health of a species and genome variability. Today this variability is manipulated to obtain a standardized set of genes, capable of imparting uniform animal size, shape and behaviour, to complete the mechanization of animal husbandry and reduction of animals to objects.

The petty engineers of life evidently know nothing about biology. Phylogenesis gives each species a genetic heritage able to express optimal solutions to environmental pressures, with wide variability to attenuate environmental fluctuations and ensure adaptation, even to the most difficult situations. Evolution is based on a balance between selective pressure and genomic variability; otherwise it would not have elements on which to make the best choice, that "selection" that gives the whole process meaning and direction. Reducing the number of species below a certain

---

[51]M. W. Head, T. J. Bunn, M. T. Bishop, V. McLoughlin, S. Lowrie, C. S. McKimmie, M. C. Williams, L. McCardle, J. MacKenzie, R. Knight, R. G. Will & J. W. Ironside, 'Prion protein heterogeneity in sporadic but not variant Creutzfeldt–Jakob disease: UK cases 1991–2002', *Ann. Neurol.* 55(6), 851–859 (2004).
[52]M. Hizume & H. Mizusawa, 'Establishment of the concept of prion diseases', *Nippon Rinsho.* 65(8), 1373–1378 (2007).
[53]R. Marchesini, *op. cit.*, p. 167, our translation.
[54]F. Jacob, *La logique du vivant. Une histoire de l'hérédité*, Gallimard, Paris (1970), our translation.

threshold or fraudulently imposing one genotype rather than another, eventually condemns life to genetic stagnation and breeds to impoverishment. The disaster awaiting us can already be glimpsed in the statistics of today:

> In the last 40 years we have lost about 2000 animal breeds of interest and today 20% of existing breeds are threatened by extinction. In Europe, the situation is alarming: 260/400 existing breeds are threatened by extinction.[55]

This is the pernicious fruit of current systems of animal production, based on inbreeding to obtain pure breeds to hand down chosen characters. Unfortunately, these animals have a similar fate to inbred humans: inbreeding produces abnormalities and loss of vigour. Accentuation of homozygosis, chromosome uniformity in certain genes or characters, goes in the opposite direction to biology, which rewards heterozygosis.

> Inbreeding goes against the natural urge for diversity typical of the living world. The logical consequence is that individuals gradually develop degenerative diseases, reproductive failure, anomalies, deformations. The reduction of genetic diversity in an individual leads to an overall deterioration of animal physiology.[56]

With the extension of biotech to animal husbandry, things can only get worse. Let us look at an example from aquaculture. Transgenic fish (such as the Japanese variety of the cyprinodont medaka, *Oryzias latipes*) modified for enhanced reproductive capacity, may lead to rapid extinction of both the engineered and the original species:

> Combining the effects of the transgene on mating success and offspring viability is predicted to result in the local extinction of any wild-type population invaded by transgenic organisms. The male mating advantage would act to increase the frequency of the transgene in the population; however, the viability disadvantage suffered by all offspring carrying the transgene would reduce the population size by 50% in less than six generations and completely eliminate the population in about 40 generations. These population projections result because the males that produce the least fit offspring obtain a disproportionate share of the mating. We refer to this type of extinction as the "Trojan gene effect", because the mating advantage provides a mechanism for the transgene to enter and spread in a population, and the viability reduction eventually results in population extinction. Such a conflict between offspring viability and male mating advantage based on large body size has been theorized to be one of the processes that can cause species extinction [...]. Such risks

---

[55] R. Marchesini, *op. cit.*, p. 174.
[56] R. Marchesini, *op. cit.*, p. 175.

should be evaluated with each new transgenic animal before release.[57]

The advantage bestowed by the transgene in this case is weight increase and enhanced reproduction. The gene is introduced into the natural population like a Trojan horse (Trojan gene hypothesis), enabling the modified species to spread the character and replace the wild-type (within 6–40 generations). Dramatically, when the acquired advantage is a weight increase, it is accompanied by reduced survival of offspring in most species studied. This ill-omened association is conceived as a major factor in the extinction of species:[58] the primitive variety becomes extinct by substitution, whereas the transgenic species, unable to compensate the high mortality of its fry (25–50%) also ends up becoming extinct. Since horizontal gene transfer has also been observed among fish, the new genes may be able to migrate from one population to another, complicating the already dismal picture.[59] The possibility of gene invasion with transmission of characters to other varieties is presumably high, with a probability of the order of 10% that the GMO variety become invasive.[60] Solutions involving sterile GM aquatic varieties are being studied to reduce the probability of gene contamination. In this way, the hypothetical advantages are of course lost along the way.[61] Different scenarios have been postulated and all underline the fact that the presence of GM fish together with non-modified fish in the same ecosystem would lead to extinction of the latter, in one way or another.[62] Things would be no better for GM varieties unable to adapt to the environment and showing handicaps, such as incapacity to avoid predators or to swim fast.[63] Conclusion: "the exploitation of GM fish in aquaculture will require, at least for the short and medium term, the use of sterile strains or of effective physical containment. Such constraints will certainly limit the economic attractiveness of GM fish".[64]

---

[57] W. M. Muir & R. D. Howard, 'Possible ecological risks of transgenic organism release when transgenes affect mating success: sexual selection and the Trojan gene hypothesis', *Proc. Natl Acad. Sci. U. S. A.* 96(24), 13853–13856 (1999).

[58] R. Lande, 'Species extinction', *Evolution* 34, 292–305 (1980).

[59] A. Koga, A. Shimada, A. Shima, M. Sakaizumi, H. Tachida & H. Hori, 'Evidence for recent invasion of the medaka fish genome by the Tol2 transposable element', *Genetics* 155, 273–281 (2000).

[60] M. Williamson, *Biological Invasions (Population and Community Biology, Series Vol. 15)*, Chapman & Hall, London (1996).

[61] S. A. Razak, G. Hwang, M. A. Rahman & N. McLean, 'Growth performance and gonadal development of growth enhanced transgenic tilapia *Oreochromis niloticus* following heat-shock induced triploidy', *Mar. Biotechnol.* 1, 533–544 (1999).

[62] P. W. Hendrick, 'Invasion of transgenes from salmon or other genetically modified organisms into natural populations', *Can. J. Fish Aquat. Sci.* 58, 841–844 (2001).

[63] A. P. Farrell, W. Bennet & R. H. Devlin, 'Growth-enhanced transgenic salmon can be inferior swimmers', *Can. J. Zool.* 75, 335–337 (1997).

[64] N. McLean, 'Genetically modified fish and their effects on food quality and human health and nutrition', *Trends Food Sci. Technol.* 14, 242–252 (2003).

Proponents of GMOs sustain that engineering of fish will in the long term make it possible to avoid many problems of current aquaculture techniques: antibiotics, adjuvants, vaccines and vitamin precursors (carotene). These admissions confirm the unnatural character of the modern food industry that does not give priority to food safety and accepts a decline in the organoleptic properties of food. Food is not only the sum of chemical ingredients, but has flavour and history; it is an aspect of the different cultures of the world and a universal human pleasure. By subtracting this quality, food is rendered substantially non-equivalent to the food offered by Mother Nature. Indeed, flavour assessment has become one of the tests currently performed on cultivated fish.[65,66]

Unfortunately, proponents of modified fish forget that GM varieties are not without their problems and risks. A multitude of unexpected effects are associated with the growth and behaviour of the fish that have so far been modified genetically.[67] Besides, gene coding for antibiotic resistance, beta-galactosidase, luciferase or chloramphenicol-acetyl transferase pose a threat for health, and "it would be correct to avoid their use in GM fish raised for human consumption".[47]

We appreciate this! Also because in this field new experiments are the last thing we need: the damage already done is evident to everyone. Any new reduction in genetic variability can only reflect negatively on the quality, stability and survival of the races we *still* have. Artificial insemination, embryo transfer, genetic manipulation to standardize sets of "perfected" genes will only increase selective pressure, directing it towards the constitution of a sort of common genome, extremely unstable and exposed to tremendous risks.

> Genetic uniformity will gather speed in an astounding manner. Soon parental uniformity will reach extremely dangerous levels because the drastic reduction in population diversity will be added to individual homozygosis. We will have an enormous quantity of farm animals with about zero genetic diversity. These two genetic malpractices will in other words lead to progressive genetic erosion within populations, which is a permanent loss! [...] What humans are doing is contrary to any logic but it is certainly in line with the trend in consumerism and with total irresponsibility towards future generations [...] First we saw a reduction in the number of breeds. Autochthonous breeds were replaced by two or three breeds with superspecific vocations (uniformity of breed). Then there was a rush for uniformity within breeds by giving the reproductive task to very few breeders (parental uniformity). This process has not yet run its course and

---

[65]E. Entis, 'Taste testing at a top Canadian restaurant', *Aqua Bounty Farms* 1, 1–4 (1998).
[66]I. Guillen, J. Berlanga, C. M. Valenzuela, A. Morales, J. Toledo, M. P. Estrada, P. Puentes, O. Hayes & J. de la Fuente, 'Safety evaluation of transgenic *Tilapia* with accelerated growth', *Mar. Biotechnol.* 1, 2–14 (1999).
[67]T. A. Ostenfeld, E. McLean & R. H. Devlin, 'Transgenesis change body and head shape in Pacific salmon', *J. Fish Biol.* 52, 850–854 (1998).

> cloning is advocated, as if parental affinity were not enough!
> Destruction of the genetic heritage of domestic animals can only
> lead to their complete elimination.[68]

Animals Fed with GMOs

Around 38 million tons of soymeal, which mostly goes into animal feed, is imported into Europe each year. Around 50–65% of this is GM or GM-contaminated, with 14–19 million tons GM-free. Food products from GM-fed animals do not have to carry a GM label. This is based on assumptions including: (1) GM DNA does not survive the animal's digestive process, (2) GM-fed animals are no different from animals raised on non-GM feed and (3) meat, fish, eggs and milk from animals raised on GM feed are no different from products from animals raised on non-GM feed. However, some of the few studies carried out in the field show that differences can be found in animals raised on *Roundup* soy animal feed, with respect to animals raised on non-GM feed, and that GM DNA can be detected in the milk and body tissues (meat) of such animals. Indeed, orally ingested foreign DNA is not completely degraded in the gut but is found in organs, blood and even the offspring of mice.[69] This finding has been confirmed both for modified soy and maize: GM DNA from GM maize and GM soy was found in milk from animals raised on these GM crops. The GM DNA was not destroyed by pasteurization.[70] GM DNA from soy was also found in the blood, organs and milk of goats. An enzyme, lactic dehydrogenase, was found at significantly raised levels in the heart, muscle and kidneys of kids fed GM RR soy.[71] This enzyme leaks from damaged cells and can indicate inflammatory or other cellular injury.

Animals Are No Longer Human

The use of cloned and/or genetically modified animals, is still controversial, though it is denied that there could be any concern about safety.[72] In the USA, a programme

---

[68]R. Marchesini, *op. cit.*, p. 176 *et seq.*

[69]R. Schubbert, U. Hohlweg, D. Renz & W. Doerfler, 'On the fate of orally ingested foreign DNA in mice: chromosomal association and placental transmission to the fetus', *Mol. Genet. Genom.* 259, 569–576 (1998).

[70]A. Agodi, M. Barchitta, A. Grillo & S. Sciacca, 'Detection of genetically modified DNA sequences in milk from the Italian market', *Int. J. Hyg. Environ. Health* 209, 81–88 (2006).

[71]R. Tudisco, V. Mastellone, M. I. Cutrignelli, P. Lombardi, F. Bovera, N. Mirabella, G. Piccolo, S. Calabro, L. Avallone & F. Infascelli, 'Fate of transgenic DNA and evaluation of metabolic effects in goats fed genetically modified soybean and in their offsprings', *Animal* 4, 1662–1671 (2010).

[72]X. C. Tian, C. Kubota, K. Sakashita, R. Okano, S. Andrew & X. Yang, 'Meat and milk compositions of bovine clones', *Proc. Natl Acad. Sci. U. S. A.* 102(18), 6261–6266 (2005).

to evaluate risk was launched by the FDA[73] after incredible delay and uncertainty, but disputes and doubts about the ethics of the whole operation involve increasingly large sectors of the scientific community, not to mention ordinary citizens: the FDA does not listen to the population at large, for whom the safety of genetically modified animals is not the real problem. The important questions are ethical ones, regarding the right of consumers to know.[74] The impact of bioengineering on the complex equilibria of Nature therefore continues to be important.

This is particularly true for the creation of "new species" of animals, that sensible people label as monsters. When speaking on this subject in a scientific context, especially in situations where economics dominates, the new sorcerers of molecular biology present their creations as revolutionary, capable of overcoming the "limitations" [sic!] of evolution, of evading the constraints that traditional techniques place on procreation and selection to obtain results with unlimited commercial uses and unlimited profits. However, when the question is posed in ecological terms to consumers' associations or competent ecological biologists, the sorcerers tell another story: bioengineers avoid terms such as "revolution" or "programming of evolutionary lineages" and state that their tools are substantially the same as those that *Homo faber* has used for thousands of years, only a tiny step ahead of hybridization, insemination and selection. These two versions are incompatible: either bioengineering produces revolutionary results that invalidate the basic mechanisms of biodiversity, evolution and natural selection, in which case it is objectively dangerous because it upsets equilibria about which we know relatively little, or it does not operate any substantial changes with respect to traditional techniques, in which case it is impossible to understand the interest and investments it is attracting. Where does the truth lie? Probably in the middle: production of GMOs truly threatens to upset ecosystems and the germplasm of the plant and animal kingdoms, but it probably will not bring any real benefits with respect to past methods that were more respectful of Nature and her laws. This is particularly true when the question is examined from the ethical point of view, focusing on the creation of "new" animals.

According to Cornelis Van Dop of Johns Hopkins Hospital in Baltimore, the crossing of animals to amplify or eliminate certain characters, has been conducted by humans since prehistoric times and has irreversibly modified the gene pools of many living species. He observes that current bioengineering techniques are about to succeed in modifying a gene at a time: selective introduction of foreign genes into germinal lines is thus the logical extension of the domestication of animals.[75]

---

[73]L. Rudenko, J. C. Matheson & S. F. Sundlof, 'Animal cloning and the FDA – the risk assessment paradigm under public scrutiny', *Nat. Biotechnol.* 25(1), 39–43 (2007).

[74]P. Vermiy, 'FDA's cloning report bypasses ethics, exposes European dilemma', *Nat. Biotechnol.* 25(1), 7–8 (2007).

[75]C. Van Dop, in: *The NIH Recombinant DNA Advisory Committee, Spring Meeting*, National Institutes of Health Bethesda, Maryland Campus, MA (1985).

This argument, so dear to the biotech lobby, is clearly false: no matter how many random or planned crosses a farmer has made, he could never produce a giant super-mouse or obtain something by crossing a sheep with a goat. In nature, it is impossible to break the barriers between species and in the rare cases in which it happens, the offspring are sterile (for example the mule, a cross between a donkey and a horse). Nature does not like mix ups. Modern chimeras are the perverse product of molecular biologists' first sallies past the columns of Hercules, namely the barrier normally separating species. These degenerate sons of Darwinian evolution are reintroducing creationism into biology, by replacing the mechanisms of evolution with their demiurgical tinkering. They have the arrogance to claim that man's capacity for direct intervention on DNA is the final evolutionary act: Van Dop claims that our capacity to manipulate genomes by recombinant DNA techniques is an integral part of evolution, it *is* evolution![76]

Too bad for them that their presumption and stupidity is revealed by real scientific data! Despite so much publicity to the contrary, no transgenic animal has yet been marketed and none of the original objectives has led to satisfactory results.[77,78]

Increased animal productivity, measured exclusively in terms of weight (difficult to conceive a more materialistic criterion), was the first testing ground on which genetic modifications failed. It began with production of the super-mouse in 1981 and already some were dreaming of super-cows, super-bulls and super-sheep. Instead, it was a super-flop. The mouse was indeed large, but it had arthritis, dyspepsia, motor instability, porous bones, and to everyone's amazement, it died young, after only 2 years of suffering compared to the 5-year life expectancy of a normal rodent.[79] No other super-animal produced enjoyed good health; all were born with many disorders and diseases, ranging from sterility and inability to walk to digestive problems and early ageing.

The outcome of attempts to make animals produce special proteins of pharmacological interest through milk was no better.[80] Apart from the fact that many substances can be synthesized quite efficiently by genetically modified microbes,

---

[76] *Ibid.*

[77] M. Buiatti, L'interazione con il genoma ospite, in: *Agro-biotecnologie nel contesto italiano*, a cura di G. Monastra, INRAN, Rome (2006), pp. 33–46.

[78] D. Humpherys, K. Eggan, H. Akutsu, K. Hochedlinger, W. M. Rideout, D. Biniszkiewicz, R. Yanagimachi & R. Jaenisch, 'Epigenetic instability in ES cells and cloned mice', *Science* 293, 95–97 (2001).

[79] J. L. Edwards, F. N. Schrick, M. D. McCracken, S. R. van Amstel, F. M. Hopkins, M. G. Welborn & C. J. Davies, 'Cloning adult farm animals: a review of the possibilities and problems associated with somatic cell nuclear transfer', *Am. J. Reprod. Immunol.* 50, 113–123 (2003).

[80] M. C. Morris & S. A. Weaver, 'Minimizing harm in agricultural animal experiments in New Zealand', *J. Agric. Environ. Ethics* 16, 421–437 (2003).

making it difficult to understand the stubborn insistence on modifying goats, once again it was easier said than done. The various attempts were not crowned with success, either because the inserted gene did not express the protein in question or because it could not be inserted in the tissue or organ of interest. It is fortunate that these attempts failed, because the allergenic potential of milk is directly proportional to its protein content, especially when the proteins are different from those in normal milk.[81] Insistence on modifying this parameter, knowing that milk is prevalently consumed by children, means exposing children to a higher risk of allergies. This is substantially why the attempt by PPL Therapeutics, a bioengineering company from Virginia, failed. The company had invested heavily on Rose, a transgenic heifer expected to produce milk high in alpha-lactoalbumin.[82] The milk, destined for premature babies, ended up causing so many problems that the project was shelved.

These costly experiments failed and lacked any rational basis. They were experiments with chimeras, impossible wishes, or wishes that bring misfortune.[83]

---

[81] P. D. Buisseret, L'allergia, in: *La Nuova Immunologia*, Le Scienze Edition, Milan (1992), p. 236 *et seq.*
[82] M. Graves, 'Transgenic livestock may become biotech's cash cow', *Los Angeles Times*, 1 May 1997, p. A12.
[83] R. Marchesini, *op. cit.*, p. 137.

5

365 new diseases[1]

According to Nero Wolfe, to know a people, it was enough
to know what they ate. The rest could be deduced: culture,
philosophy, ethics, politics ...

R. Stout

What is food for one man is poison for another.

Titus Lucretius Caro

The less people know about how sausages and laws are made,
better they sleep at night.

Otto von Bismarck

We Have No More Antibiotics

In 1997, three teams of doctors on three continents lived several days of authentic terror. Patients with *Staphylococcus aureus* infections, notoriously resistant to most antibiotics, did not respond to the antibiotic of last resort, vancomycin. Fortunately they responded to a cocktail of other drugs, however the fear remains. The appearance of vancomycin-resistant strains of *Staphylococcus* is an enormous problem. This insidious major infection contracted in hospitals could become unstoppable.

We know that many strains of *Staphylococcus* in the world are resistant to all antibiotics and the same is true of other bacteria, such as certain strains of *Mycobacterium tuberculosis*, *Enterococcus faecalis* and *Pseudomonas aeruginosa*. They completely elude the hundred-odd products available to doctors. This explains why deaths due to contagious diseases (such as TB), once regarded as overcome, are now increasing in industrialized countries.[2]

How did this happen? The many reasons would fill a volume, an interesting volume that would explain how we are poisoning ourselves with the fruits of "progress". The heart of the problem is so-called *resistance*, a complex phenomenon

---

[1]The title of the chapter is from J. Dormont, O. Bletry & J. F. Delfraissy, Les *365 Nouvelles Maladies*, Flammarion Médecine, Sciences, Paris (1989).
[2]M. Bizzarri, Epidemiologia e multiculturalismo, in: *Medicina e Multiculturalismo*, S. Maffettone (Ed.), Questioni di Bioetica, Apeiron, Bologna (2000), p. 155.

caused by various mechanisms by which a strain of bacteria arms itself with defences that make drugs inoffensive or that render the strain insensitive to drugs.

The list of microbes that respond poorly or not at all to available drugs is long and getting longer. It includes *Staphylococcus aureus, Acinetobacter, Enterococcus faecalis, Neisseria gonorrheae, Haemophilus influenzae, Mycobacterium tuberculosis, Escherichia coli, Pseudomonas aeruginosa, Shigella dissenteriae* and *Streptococcus pneumoniae*. Resistance may be natural or acquired: in the first case the germ is constitutionally insensitive to the drug. Acquired resistance may be due to genetic mutations or modifications caused by incorporation of new genetic material: the donor may be a bacterium, a virus or even a food.[3,4]

The process by which resistant populations of bacteria are selected is simple. When a colony of microbes comes into contact with an antibiotic, most of the bacteria sensitive to the drug die. The few bacteria with natural or acquired resistance to the drug are free to proliferate without competition for nutrients. The effect is accentuated if the colony is treated with insufficient doses of the drug, which do not even kill all sensitive bacteria. The latter then conjugate, transduce and transform, exchanging DNA segments that impart resistance. Abuse of antibiotics, taking them in wrong doses and without any clear aim, has other negative consequences, such as elimination of inoffensive microbes and saprophytes that prevent pathogens from establishing and acting. Normally harmless bacteria, resistant to drugs, may also be spared, but their spread increases the bank of characters of resistance in the bacterial population, thus increasing the probability that the characteristic be transmitted to pathogenic bacteria. Even a normal saprophyte may become a killer in this way. This is demonstrated by the story of *Pseudomonas aeruginosa*, once a peaceful guest of the human body, today public enemy number one in hospitals all over the world.

The problem of resistance to antibiotics does not seem to have raised the interest it merits in the scientific community or health authorities, despite its dramatic nature and the anxiety it is causing specialists in infectious diseases. The dominant attitude is faith that a solution will be found and that there is no hurry to address the problem. This attitude may not surprise the worldly, as it is a good example of the superficiality with which complex operations are tackled today. The repercussions on ecosystems are unknown and will probably be unpredictable for a long time to come. But what has this to do with transgenic foods? The same question is asked by molecular biologists, ignorant of the epidemiology of infectious diseases, who happily traffic with viruses and bacteria, unaware of the risk they pose to the human race.

---

[3]H. H. Wang, M. Y. Manuzon, M. Lehman, K. Wan, H. Luo, T. E. Wittum, A. E. Yousef & L. O. Bakaletz, 'Food commensal microbes as a potentially important avenue in transmitting antibiotic resistance genes', *FEMS Microbiol. Lett.* 254, 226–231 (2006).
[4]C. Serra, 'La resistenza vien mangiando', *Le Scienze* 468, 37 (2007).

Massive and indiscriminate use of drugs has certainly accelerated the development of bacterial resistance. For example, the 1,000 tons of antibiotics produced in the USA in 1954 has risen to more than 22,000 tons! It is also true that increasing responsibility for this situation can be ascribed to intensive industrial agriculture and animal husbandry, which rely heavily on the generous systematic use of antibiotics. Nearly 50% of world antibiotic production is used on farm animals for therapeutic or production purposes,[5] usually to favour weight increase and protect animals from superinfections induced by other drugs and hormones. Such inappropriate and reckless use is certainly a cause of the emerging drug resistance shown by many pathogenic microorganisms.[6] Use of antibiotics in animal husbandry, prohibited in Europe but still widespread in the USA and other countries, is a bad practice that can have serious epidemiological consequences for animals and humans.[7,8] Thoughtless use of antibiotics in agriculture and animal husbandry is believed to have a much greater impact on the spread of bacterial resistance than clinical antibiotic use and hospitalization.[9,10]

There are many complex reasons for this hyperconsumption of antibiotics. First there is indiscriminate use of antibiotics in the countryside where they are sprayed on fruit trees to prevent or control bacterial infections. High concentrations destroy any form of microbial life, and drug residues favour the development of resistant strains, that subsequently attack the fruit, especially during processing and transport. Spray can also contaminate other plants and carry great distances, in quantities insufficient to eradicate infections but sufficient to kill sensitive bacteria, promoting further focuses of resistance. For example, it has been shown that concentrations of bacteria on the skin of family members of persons who continually use antibiotics increases sharply, even if the family members do not use the drug.[11] Thus, in environments (hospitals and animal farms) where there is high dispersion of antibiotics, accidentally or professionally, the number of bacteria resistant to one

[5]V. Perreten, F. Schwartz, L. Cresta, M. Boeglin, G. Dasen & M. Teuber, 'Antibiotic resistance spread in food', *Nature* 389, 801–802 (1997).

[6]S. L. Gorbach, 'Editorial: antimicrobial use in animal feed: time to stop', *N. Engl. J. Med.* 345(16), 1202–1203 (2001).

[7]T. L. Sorensen, M. Blom, D. L. Monnet, N. Frimodt-Muller, R. L. Poulsen, F. Espersen, 'Transient intestinal carriage after ingestion of antibiotic-resistant *Enterococcus faecium* from chicken and pork', *N. Engl. J. Med.* 345(16), 1161–1166 (2001).

[8]L. C. McDonald, S. Rossiter, C. Mackinson, Y. Y. Wang, S. Johnson, M. Sullivan, R. Sokolow, E. DeBess, L. Gilbert, J. A. Benson, B. Hill & F. J. Angulo, 'Quinupristin-dalfopristin–resistant *Enterococcus faecium* on chicken and in human stool specimens', *N. Engl. J. Med.* 345(16), 1155–1160 (2001).

[9]A. A. Salyers, 'Antibiotic resistance transfer in the mammalian intestinal tract: implications for human health, food safety and biotechnology', Springer, New York; RG Landes, Austin (1995).

[10]D. L. Smith, J. Dushoff, J. G. Morris Jr, 'Agricultural antibiotics and human health', *Plos Med.* 2(8), 1–5 (2005).

[11]S. B. Levy, 'The challenge of antibiotic resistance', *Sci. Am.* March, 46–53 (1998).

or more anti-microbic drugs increases in persons or animals that do not use the product but merely live in or pass through those environments, or live nearby! This is primarily due to the fact that the genetic characteristic that imparts resistance is more readily acquired when antibiotic concentrations are subpharmacological. It can then be transferred by plasmids from one bacterial strain to another.[12] A second cause is due to the nature of intensive animal raising that upsets the feeding habits and hygiene of livestock forced to live in unhealthy, overcrowded conditions under severe stress. Promiscuity, attenuation of immune defences and filthy conditions are other factors that facilitate the propagation and attack of bacteria. The widespread practice of treating cows with bovine growth hormone increases milk production by subjecting the udders to anti-physiological stress, exposing almost all cows to mastitis. In both cases, the diet must be supplemented with large doses of antibiotics for therapeutic and prophylactic reasons.[13] Substantial concentrations of drug residues can build up in animal flesh, withstand processing and cooking, and arriving unmodified on our plates.

These measures revealed that routine administration of low-dose antibiotics causes weight increase, though there has not yet been a full scientific explanation. Adding drugs to feed took on in a flash and has become a routine practice not subject to any control: 80% of animals raised in the USA today are treated with hormones and antibiotics. This explosive mixture has many consequences, including favouring the development of resistant bacteria.

A first worrying balance of the consequences for human health was published in 1997.[14] For example, a strain of *Salmonella typhimurium* (Dt104), responsible for human salmonella and gastroenterocolitis, is now multiresistant to almost all antibiotics. Research conducted in Scotland and Wales showed that in 95% of cases, the bacterium isolated from animals had not only become resistant to sulphamides but also to four antibiotics commonly used to treat it: ampicillin, chloramphenicol, streptomycin and tetracyclin. Almost all strains of the bacterium were resistant to ampicillin; in 80% of cases – out of 182 strains examined – the phenomenon also involved other anti-bacterial agents.[15] Strain Dt104, the natural reservoir of which was originally cattle, has now spread to chickens, sheep and pigs. The appearance of resistant strains initially discovered in England, homeland of many animal contaminations, as well as the USA, then spread to France and other European

---

[12]D. G. White, S. Zhao, R. Sudler, S. Ayer, S. Friedman, S. Chen, P. F. McDermott, D. D. Wagner & J. Meng, 'The isolation of antibiotic-resistant *Salmonella* from retail ground meats', *N. Engl. J. Med.* 345(16), 1147–1154 (2001).
[13]Pure Food Campaign, *What's Wrong with Genetically Engineered Foods?*, Washington DC (1995), p. 134.
[14]C. Gorman, 'Germ warfare. A drug resistant staph strain has doctors on edge', *Time* 1 September (1997).
[15]M. Perez, 'Lourdes menaces sur l'efficacité des antibiotiques', *Le Figaro* 18 November (1997).

countries.[16] The example of *Salmonella* is not isolated, which is why the EU decided to ban the use of new glycopeptide antibiotics in animal husbandry. Avoparcin, for example, is believed to favour cross-resistance to vancomycin, the only agent left for fighting dreadful *Staphylococcus aureus* infections.

This already dramatic situation may be aggravated by technologies used in the production of GM seeds. Indeed, to identify genes introduced into GM organisms, it is often deemed necessary to insert a "resistance to antibiotics gene" that acts as a marker. The procedure is simple: a vector is used to insert a gene coding for the required protein in a cell culture; a gene segment that gives the transfected cell resistance to one or more antibiotics is associated with the former gene. To be sure that the transplant was successful, the culture is treated with the antibiotic(s): cells that have not acquired resistance obviously die, while those that survive have evidently incorporated the resistance gene and together with it, whatever the molecular biologist aimed to modify. Sometimes more than one resistance gene is inserted: the second is inserted in the "expression vector" (that stimulates activation of the modified gene segment) when the latter is first cloned in bacterial cells to obtain many copies, before using it to transform the DNA of the cell. This procedure proved necessary because under current laboratory conditions, gene transfer is an extremely uncertain and inefficient process: on the average, gene transfer succeeds in only one cell in a thousand. The marker gene, generally a sequence that protects against an antibiotic or a herbicide, is added to identify the cells that have accepted the gene. The cells surviving treatment with the drug are then selected as homogeneous "modified" progeny. The problem is that the resistance gene, extraneous to the aim being pursued, is transmitted to subsequent generations, becoming part of the genetic heritage of the species.

The failure of a series of experimental attempts suggested that the antibiotic resistance gene could not be transferred from a modified plant to a bacterium.[17] However, the history of molecular biology is studded with relative impossibilities: what seems unlikely today can be real tomorrow. Gebhard and Smalla, researchers from *Braunschweig University*, succeeded in showing that when inserted in the genome of transgenic sugar beet, the gene *nptII* that confers resistance to kanamycin, can transfer antibiotic resistance to *Acinetobacter* sp.[18] This was later

---

[16]E. J. Threlfall, 'Incidence croissante de la résistance au triméthoprime et à la ciprofloxacine de *Salmonella typhimurium* DT 104 épidémique en Angleterre et au pays de Galles', *Eurosurveillance* 2(11), 77–81 (1997).

[17]K. Schluter, J. Futterer & I. Potrykus, 'Horizontal gene transfer from a transgenic potato line to a bacterial pathogen (*Erwinia shrysanthemi*) occurs, if at all, at an extremely low frequency', *Biotechnology* 13, 1094–1098 (1995).

[18]F. Gebhard & K. Smalla. 'Transformation of *Acinetobacter* sp. strain BD413 by transgenic sugar beet DNA', *Appl. Environ. Microbiol.* 64, 1550–1554 (1998).

also documented for other gene sequences imparting resistance to antibiotics such as neomycin and ampicillin.[19]

Bacterial strains acquiring the resistance gene in this way could obviously not be eliminated by the specific antibiotic, which would have become useless. Although it is technically possible to eliminate the gene in question within certain limits, the operation is neither simple nor economic. This led to recommendations from many quarters to remove marker genes from modified species, but this advice was not generally heeded by major producers. Varieties modified in this way can therefore be cultured commercially and introduced into animal feed and human food together with the gene for antibiotic resistance. This is clearly an alarming prospect for anyone with a few notions about infectious diseases. The practice has been criticized by scientists who are not against transgenic plants,[20,21] but producers of GM seeds defend their actions in different ways. The arguments rely on the low probability of transmission of markers of resistance to bacteria of the intestinal flora and on the statistical improbability of any consequences,[22] since the human intestine already contains high concentrations of resistant microbes. These claims are easily refuted: however infrequent an event may be, it becomes statistically significant when applied to populations of hundreds of millions of individuals. The human intestine contains billions of bacteria and potential consumers of GM foods already number hundreds of millions. A degree in mathematics is not necessary to calculate that exceptionally low frequencies can lead to a significantly high number of events of gene transmission:

> [...] The magnitude of transformation frequency for various bacterial strains recognized by visible transformed colonies is 10^{-4} to 10^{-8}, and up to 10^{-16} in the absence of such colonies. The number of symbiotic bacteria per gram of the contents of intestines is as high as 10^{11}. When this number is calculated per total intestinal contents, the probability of the transformation of symbiotic bacteria becomes quite high.[23]

The probability also increases considerably under selective pressure, or in persons undergoing antibiotic therapy. Once even a single microbe acquires genetically

[19] M. Droge, A. Puhler & W. Selbitschka, 'Horizontal gene transfer as a biosafety issue: a natural phenomenon of public concern. *J. Biotechnol.* 64, 75–90 (1998).

[20] H. A. Kupier & G. A. Kleter, 'The scientific basis for risk assessment and regulation of genetically modified foods', *Trends Food Sci. Technol.* 14, 277–293 (2003).

[21] P. M. Bennett, C. T. Livesey & D. Nathwani, 'An assessment of the risks associated with the use of antibiotic resistance genes in genetically modified plants: report of the Working Party of the British Society for Antimicrobial Chemotherapy', *J. Antimicrob. Chemother.* 53, 418–431 (2004).

[22] A. L. Konov, 'Biotechnology and the horizontal gene carrying: can one eat genetically modified products and acquire the resistance to antibiotics?', *Ekologiya I Zhizn'* 2, 66–68 (2002).

[23] A. M. Kulikov, 'Genetically modified organisms and risks of their introduction', *Russian J. Plant Physiol.* 52(1), 99–111 (2005).

advantageous information, under conditions in which natural selection is acting, this information can spread quickly to other bacterial populations by horizontal gene transfer.[24] Under such conditions, and especially when non-optimal doses of the drug are used, non-resistant bacteria are eliminated and the stage is set for the survival of bacteria that have acquired genes conferring resistance to that antibiotic.[25]

It is therefore inappropriate to play down the importance of the phenomenon by saying that it only involves rarely used antibiotics, such as kanamycin and neomycin,[26] to which many microbial strains have already developed resistance. Kanamycin and neomycin are still part of the therapeutic battery of doctors and are specific for certain infections. For this very reason, anything promoting bacterial resistance should be discouraged, and resistance could also be fought by using antibiotics more specifically and less thoughtlessly.[27] Secondly, the phenomenon of resistance and its horizontal transfer also involve other currently used broad-spectrum antibiotics, such as ampicillin. To argue only about kanamycin is a strategy for creating confusion and hiding other truths.

In 2005, Syngenta, a leader in the food biotech sector, admitted that Bt11 maize marketed in many nations had been inadvertently contaminated with another modified species of maize, Bt10, in the period 2001–2004. Bt10 contains the gene *bla* that confers resistance to ampicillin as marker and was not authorized for human consumption by the FDA. The incident was considered alarming for various reasons. First, because like many other episodes in recent years, it revealed the inadequacy of measures taken by producers and controlling bodies to punctually ensure that transgenic seeds released on the market were of approved types. Controls are insufficient and it is probably legitimate to believe that they are not conducted in an exemplary manner. There are objectively too many possibilities of accidental (really?) contamination to be effectively monitored. The production line from origin to consumer is long and often does not ensure absolute separation of the different modified and unmodified species. Moreover, only thanks to a communication by the Advisory Committee on Releases to the Environment (Great Britain) did it become

---

[24]G. Van der Eede, H. Aarts, H. J. Buhk, G. Corthier, H. J. Flint, W. Hammes, B. Jacobsen, T. Midtvedt, J. van der Vossen, A. von Wright, W. Wackernagel & A. Wilcks, 'The relevance of gene transfer to the safety of food and feed derived from genetically modified (GM) plants', *Food Chem. Toxicol.* 42, 1127–1156 (2004).
[25]T. R. Licht, C. Struve, B. B. Christensen, R. L. Poulsen, S. Molin & K. A. Krogfelt, 'Evidence of increased spread and establishment of plasmid RP4 in the intestine under sub-inhibitory tetracycline concentrations', *Fems Microbiol. Ecol.* 44, 217–223 (2003).
[26]G. D'Agnolo, 'GMO: human health risk assessment', *Vet. Res. Comm.* 29(Suppl. 2), 7–11 (2005).
[27]In Finland they have begun to adopt such measures: with fewer prescriptions for erythromycin, the frequency of strains of resistant *S. pyogenes* dropped from 19.8% to 0% in no time (cf. S. D. Levy, *Antibiotic Paradox: How Miracle Drugs Are Destroying the Miracle*, Plenum, New York (1992), p. 172).

public that Bt10 maize contained resistance genes to a widely used, broad-spectrum antibiotic such as ampicillin. Neither Syngenta nor the US EPA deemed it necessary to mention this when Bt10 was approved: "It is quite scandalous [...]. This shows that the government and the company are not being forthright" was the comment of Greg Jaffe, director of the biotech project at the *Center for Science in the Public Interest*.[28] After a long and deafening silence, the authorities tried to deny the facts, but then had to admit the evidence. This they did through clenched teeth, forced by revelations and enquiries conducted by the prestigious journal *Nature*.[29,30] Then they set to work devising replies that convinced nobody, first saying that presence of the gene was irrelevant to health and for discussions about safety, since the risk of horizontal transmission was "very low", subsequently admitting candidly that they knew what the company had told them about the gene and that they could not discuss that aspect of the problem at the time. Margaret Mellon's considered this terrible,[30] and, *Nature* that had followed the affair closely, condemned the behaviour of the US government regulatory bodies, declaring that they were not to be trusted because they were unconscious towards their task.[29] After responsibility had been passed back and forth between the Department of Agriculture, FDA and EPA, *Nature* claimed that the Environmental Protection Agency had nothing to do with the environment and was certainly not protecting anyone. For 4 years, Bt11 maize was sold mixed with Bt10 and Syngenta only realized the error by chance, after a subsidiary reported the contamination. Nevertheless, after months of lengthy negotiations with EPA to solve the problem, Syngenta still refused to provide the list of countries that had accidentally received Bt10 seeds. This not only demonstrates the lack of an overall system for monitoring GMO, raising doubts about the adequacy of controls,[31] but raises legitimate doubts that controls are actually possible. A basic doubt remains:

> [...] Syngenta has attributed the belated discovery of its inadvertent release of Bt10 to progress in the technology that it uses to monitor seeds. If true, this explanation will come as news to anyone who had assumed that the agricultural-biotechnology industry had known from the start what transgenes it was putting into its seeds. If the discovery was simply a matter of happenstance, we can have little confidence that similar problems won't occur again.[32]

Although the ampicillin-resistance gene does not seem to be a problem for EPA or Syngenta, it worries the Europeans, because many major strains of pathogens are

---

[28]C. Macilwain, 'Stray seeds had antibiotic-resistance genes', *Nature* 434, 548 (2005).
[29]'Don't rely on Uncle Sam', *Nature* 434, 807 (2005).
[30]C. Macilwain, *op. cit.*
[31]C. Macilwain, 'US launches probe into sales of unapproved transgenic corn', *Nature* 434, 423 (2005).
[32]Don't rely on Uncle Sam, *op. cit.*

still fortunately sensitive to ampicillin and "occasional transfer of these particular resistance genes from GM plants to bacteria would pose an unacceptable risk to human or animal health."[33] Gene transfer may be relatively insignificant but take on great importance if it ensures a selective advantage.[34] Moreover, if the cause of gene transfer spreads resistance, it will mean that other drugs will have to be used to treat animal diseases and this will have big consequences for consumers throughout the food chain, especially considering that the gene construct inserted in Bt10 maize has structural characteristics that make it extremely efficient in ensuring gene transfer.[35] The considerations do not convince those in favour of GMOs but they led to banning of antibiotic resistance markers in Europe (Dir 2001/18/EC), where systems to remove gene markers from modified seeds[36] once plant DNA is transformed is now obligatory. Such systems have been known for some time[37] but, inexplicably, never used.

With these preliminaries, transgenic foods, still limited to a few nations, could become much more than an unconfirmed source of resistance to antibiotics. We saw that the gene markers most often used impart ampicillin and kanamycin resistance (genes *bla* and *nptII*). The first is a front-line antibiotic used in medicine, the second is still important, though seldom used in the usual therapeutic protocols, and has affinities with other antibiotics, so that kanamycin resistance could extend to other aminoglucosides such as tobramycin and amikacin that are used clinically and are often determinant for curing sepsis and otherwise untreatable infections. These cross-resistance phenomena are well known to microbiologists and as far as kanamycin is concerned, have been demonstrated in certain strains

---

[33]K. L. Goodyear, 'Comment on: an assessment of the risks associated with the use of antibiotic resistance genes in genetically modified plants: report of the Working Party of the British Society for Antimicrobial Chemotherapy', *J. Antimicrob. Chemother.* 54(5), 959 (2004). The italics are mine.

[34]M. J. Gasson, 'Gene transfer from genetically modified food', *Curr. Opin. Biotechnol.* 11, 505–508 (2000).

[35]G. Azeez, 'Ampicillin threat leads to wider transgene concern', *Nature* 435, 561 (2005).

[36]It is currently uncertain whether such technology is effective and safe (cf. G. Van der Eede, H. Aarts, H. J. Buhk, G. Corthier, H. J. Flint, W. Hammes, B. Jacobsen, T. Midtvedt, J. van der Vossen, A. von Wright, W. Wackernagel & A. Wilcks, *op. cit.*). *Novartis* was among the first companies to remove antibiotic resistance markers and developed an alternative system (*Positech*) based on selective growth of modified seeds (phospho-mannose isomerase gene insert) in the presence of mannose ('Novartis pins hopes for GM seeds on new marker system', *Nature* 406, 924 (2000)). Other mechanisms are also now available (T. Komari, Y. Hiei, Y. Saito, N. Murai & T. Kumashiro, 'Vectors carrying two separate T-DNAs for co-transformation of higher plants mediated by *Agrobacterium tumefaciens* and segregation of the transformants feed from selection marker', *Plant J.* 10, 165–174 (1996)), however the problem posed by antibiotic resistance genes is still topical and a source of concern.

[37]E. C. Dale & D. W. Ow, 'Gene transfer with subsequent removal of the selection gene from the host genome', *Proc. Natl Acad. Sci. U. S. A.* 88, 10558–10562 (1991).

of *Bacillus subtilis*:[38] their importance will certainly increase, with predictably disastrous repercussions.

Antibiotic resistance is a very serious problem that we should all be concerned about. Because of the increasing spread of antibiotic resistance and the lack of valid alternative antibiotics, microbiologists all over the world are tirelessly urging regulatory bodies to block all routes by which resistance can spread, especially those between animals, food and consumers. A traumatic return to pre-antibiotic days, the spread of emerging infectious diseases and the reappearance of old diseases that have now become almost insensitive to drugs are the price to be paid for not acting. The dramatic importance of this alarm does not yet seem to have been understood. Some are still back in 1969, when victories over infectious diseases came quick and fast, and it was thought that certain diseases would be completely eliminated. In those days, a prestigious US medical authority, William H. Stewart, could lightly claim that infectious diseases were dead and buried. But he underestimated the adversary and the force of natural selection. The naked truth is that pathogens can adapt to any substance researchers develop. A scientist recently commented that the war had been won ... but by the adversary.[39] We should prepare to raise the white flag.

Gene Transfer

With great apprehension, biologists have long suspected the existence of the so-called horizontal transmission of genes, a phenomenon by which DNA segments found in a certain microenvironment are absorbed by bacteria as part of their chromosome heritage. It should first be said that this process regards DNA fragments from any source (not necessarily transgenic) and that it becomes important for human health when it occurs in the intestines, between undigested DNA residues from food and microbe populations.

Gene transfer has been observed in cultures[40] and in organic microenvironments of animals *in vivo*: in chickens[41] and mice,[42] in the digestive tract and

---

[38] F. Saldman, *Les Nouveaux Risques Alimentaires*, Ramsay, Paris (1997).

[39] R. M. Nesse & G. C. Williams, 'Evolution and origin of disease', *Sci. Am.* November (1998).

[40] J. R. Saunders *et al.*, Genotypic and phenotypic methods for the detection of specific released microorganisms, in: C. Edwards (Ed.), *Monitoring Genetically Modified Organisms in the Environment*, Wiley, New York (1993), p. 27.

[41] J. F. Guillot, *In vivo* transfer of a conjugative plasmid between isogenic *Escherichia coli*, strains in the gut of chickens, in the presence and absence of selective pressure, in: M. J. Gauthier (Ed.), *Microbial Releases*, Springer Verlag, Berlin (1992), pp. 123–130.

[42] A. Wilcks, A. H. A. M. van Hoek, R. G. Joosten, B. B. L. Jacobsen & H. J. M. Aarts, 'Persistence of DNA studied in different *ex vivo* and *in vivo* rat models simulating the human gut situation', *Food Chem Toxicol.* 42, 493–502 (2004).

mouth.[43] The prerequisite for transfer was that the segments of nucleic acid survived the acid environment of the stomach and reached the intestines, where symbiotic bacterial flora is concentrated, in significant quantities and lengths.

Until a few years ago, it was thought that DNA in food was completely broken down by the combined action of pancreatic nucleases, enzymes and stomach acid. Now we know that this breakdown is neither complete nor instantaneous and may be considerably modulated by different factors (length of the intestine, low acid content, chronic constipation) that prolong the survival of extraneous DNA, favouring slow intestinal transit.[44] Whole gene sequences can survive cooking and processing of food; the usual procedure used to treat meat and bone meal and other foods (90°C dry heat for 30 min) leaves DNA practically intact and in concentrations greater than 100 parts per billion (ppb), well above the 25 ppb threshold necessary to avoid the development of autoimmune reactions and horizontal transfection.[45]

Studies conducted by the Institute of Genetics of Cologne University showed that a significant percentage (5%) of extraneous DNA sequences (from the phage M13) added to the diet of mice, passed the gastric barrier and were absorbed by the circulation, reaching distant organs and cells. The DNA was detected in leucocytes, intestinal Peyer patch cells, spleen and kidney cells, where it became a stable part of the host-cell genome.[46,47] Fragments of the same DNA sequences fed to pregnant mice passed the placenta and reached the foetus, where they were captured by foetal organs and cells.[48] A study conducted directly on humans showed foreign (rabbit) mitochondrial DNA in blood of volunteers who ate rabbit meat.[49] These studies raised important questions and concerns: insertion of foreign DNA (especially viral DNA) in the genome of eukaryotic cells of mammals triggers a series of complex events that tend to silence the activity of the transfected gene segment in order to

[43]F. Doucet, Conjugal transfer of genetic information in gnotobiotic mice, in: M. J. Gauthier (Ed.), *Microbial Releases*, Springer Verlag, Berlin (1992), p. 58.

[44]M. Palka-Santini, B. Schwarz-Herzke, M. Hosel, M. Renz, S. Auerochs, H. Brondke & W. Doerfler, 'The gastrointestinal tract as the portal of entry for foreign macromolecules: fate of DNA and proteins', *Mol. Gen. Genomics* 270, 201–215 (2003).

[45]J. M. Forbes, 'Effect of feed processing conditions on DNA fragmentation', UK Ministry of Agriculture Fisheries and Food, section 5 scientific report, London (1998).

[46]R. Schubbert, C. Lettmann & W. Doerfler, 'Ingested foreign (phage M13) DNA survives transiently in the gastrointestinal tract and enters the blood-streams of mice, *Mol. Gen. Genet.* 242, 495 (1994).

[47]R. Schubbert, D. Renz, B. Schmitz & W. Doerfler, 'Foreign (M13) DNA ingested by mice reaches peripheral leukocytes, spleen, and liver via the intestinal wall mucosa and can be covalently linked to mouse DNA', *Proc. Natl Acad. Sci. U. S. A.* 94, 961–966 (1997).

[48]R. Scubbert, U. Hohlweg, D. Renz & W. Doerfler, 'On the fate of food ingested foreign DNA in mice chromosomal association and placental transmission to the foetus,' *Mol. Gen. Genet.* 259, 569–576 (1998).

[49]A. Forsman, D. Ushameckis, A. Bindra, Z. Yun & J. Blomberg, 'Uptake of amplifiable fragments of retrotrasposon DANN from the human alimentary tract', *Mol. Gen. Genomics* 270, 362–368 (2003).

maintain the integrity and identity characteristics of the host genome.[50] However, if this process is inefficient, insertion of foreign DNA could play a non-negligible role in arousing evolutionary or pathogenetic phenomena, including growth of tumours and accentuation of mutagenic processes.[51] W. Doerfler pointed out the risks associated with insertion of viral DNA in the human genome, describing a fascinating and at the same time terrifying picture. Insertion of DNA fragments associated with parts of a virus vector may lead to insertional mutagenesis,[52] in the broadest sense. Changes in genome structure, levels of methylation of DNA and consequent alterations in protein expression are not necessarily limited to the area of insertion of the foreign DNA, but may affect different regions and upset the stability of the whole filament. These epigenetic modifications can be just as powerful and dangerous as true mutations.[53] Our bodies can obviously incorporate foreign DNA through diet, especially if DNA fragments are associated with viral elements having the capacity to overcome cell membrane defences. Indeed, modified genes are associated with viral vectors to exploit their penetration capacity and reduce the probability that the foreign gene be rejected by the recipient cell. Insertional mutagenesis is a new route for viruses to conduct oncogenic activity and this obviously has extraordinary repercussions for the interpretation of the results obtained working with transgenic organisms and foods. At the same time, it poses threats that cannot be underestimated, especially considering the plague proportions of cancer and malignant diseases in our society.

As it was reasonable to expect, survival in the gastrointestinal tract has also been confirmed for transgenic DNA segments in animals fed GM foods. The probability of this occurring depends on many factors involving DNA stability and the number of copies of the transgene inserted in the modified food.[54,55] Certain studies[56,57]

---

[50]H. Heller, C. Kammer, P. Wilgenbus & W. Doerfler, 'Chromosomal insertion of foreign (adenovirus type 12 plasmid or bacteriophage) DNA is associated with enhanced methylation of cellular segments', *Proc. Natl Acad. Sci. U. S. A.* 92, 5515–5519 (1995).

[51]W. Doerfler, R. Schubbert, H. Heller, C. Kammer, K. Hilgher-Eversheim, M. Knoblauch & R. Remus, 'Integration of foreign DANN and its consequences in mammalian systems', *Tibtech* 15, 297–301 (1997).

[52]W. Doerfler, 'A new concept in adenoviral oncogenesis: integration of foreign DNA and its consequences', *Biochim. Biophys. Acta* 1288, F79–F99 (1996).

[53]R. Holliday, 'Viral mutagenesis and insertional mutagenesis', *Science* 238, 163 (1987).

[54]A. Klotz, J. Mayer & R. Eispanier, 'Degradation and possible carry over effects of feed DNA monitored in pigs and poultry', *Eur. Food Res. Technol.* 214, 271–275 (2002).

[55]T. Reuter & K. Aulrich, 'Investigation on genetically modified maize (Bt maize) in pig nutrition: fate of feed-ingested foreign DNA in pig bodies', *Eur. Food Res. Technol.* 216, 185–192 (2003).

[56]A. Einspanier, A. Klotz, A. Kraft, K. Aulrich, R. Poser, G. Jahreis & G. Flachowsky, 'The fate of forage plant DNA in farm animals: a collaborative case-study investigating cattle and chicken fed recombinant plant material', *Eur. Food Res. Technol.* 212, 129–134 (2001).

[57]R. H. Phipps, E. R. Deaville & B. C. Maddison, 'Detection of transgenic and endogenous plant DNA in rumen fluid, duodenal digesta, milk, blood and feces of lactating dairy cows. *J. Dairy Sci.* 86, 4070–4078 (2003).

failed to find traces of the transgenic segment in biological fluids and tissues of animals fed GM feed, though DNA sequences from the plant chloroplast genome were detected; other studies documented the presence of transgenic segments in blood, liver, spleen and kidneys of the animals.[58,59,60]

This phenomenon is not limited to animals fed with GM feed, but also regards humans. Six to twenty-five per cent of GM plasmids survive degradation by saliva enzymes and become incorporated in the DNA of *Streptococcus gordonii*, a normal saprophyte of the human mucous membranes.[61] The transgene *bla* of GM maize, that codes for beta-lactamase, the enzyme providing ampicillin resistance, resists enzymatic breakdown for about an hour when incubated in saliva.[62] The frequency with which transformation occurs is proportional to the time spent in the mouth, increasing sharply after 10 min. Saliva probably contains factors that facilitate genetic exchange, since this process does not occur in synthetic solutions but has also been found active in studies with sheep.[63] Since this phenomenon not only regards sequences coding for antibiotic resistance but any DNA segment not fully broken down by food processing, spontaneous genetic recombination processes may create bacterial mosaics with new characteristics, not all of which favour human health. DNA surviving passage through the whole gastrointestinal tract seems to conserve its transforming capacity, namely the capacity to transmit its specific characteristics, including antibiotic resistance, to other cells and organisms. In a study on gnotobiotic rats (raised under controlled conditions so that their intestines could only be colonized by certain bacteria), Wilcks and colleagues[64] found not only that plasmid DNA administered in feed appeared throughout the gastrointestinal tract, but it also maintained the capacity to transform bacteria by transfer of genetic information. Similarly, in simulated human

[58]R. Sharma, D. Damgaard, I. W. Alexander, M. E. R. Dugan, J. L. Aalhus, K. Standford & T. A. McAllister, 'Detection of transgenic and endogenous plant DNA in digesta and tissues of sheep and pigs fed Roundup Ready canola meal', *J. Agric. Food Chem.* 54, 1699–1709 (2006).

[59]E. H. Chowdhury, O. Mikami, Y. Nakajima, A. Hino, H. Kuribara, K. Suga, M. Hanazumi & C. Yomemochi, 'Detection of genetically modified maize DNA fragments in the intestinal contents of pigs fed Star-Link CBH351', *Vet. Hum. Toxicol.* 45, 95–96 (2003).

[60]R. Mazza, M. Soave, M. Morlacchini, G. Piva & A. Morocco, 'Assessing the transfer of genetically modified DNA from feed to animal tissues', *Transgenic Res.* 14, 775–784 (2005).

[61]D. K. Mercer *et al.*, 'Fate of free DNA and transformation of the oral *Streptococcus gordonii* DL1 by plasmid DNA in human saliva', *Appl. Environ. Microbiol.* 65, 66 (1999).

[62]P. S. Duggan, P. A. Chambers, J. Heritage & J. M. Forbes, 'Survival of free DNA encoding antibiotic resistance from transgenic maize and the transformation activity of DNA in ovine saliva, ovine rumen fluid and silage effluent', *FEMS Microbiol. Lett.* 191, 71–77 (2000).

[63]P. S. Duggan, P. A. Chambers, J. Heritage & J. M. Forbes, 'Fate of genetically modified maize DNA in the oral cavity and rumen of sheep', *Br. J. Nutr.* 89, 159–166 (2003).

[64]A. Wilks, A. H. A. M. van Hoek, R. M. Joosten, B. B. L. Jacobsen & H. J. M. *op. cit.*

studies it was found that transgenic sequences resisted degradation by saliva and digestive enzymes, entering the intestines.[65] A similar study was conducted for the first time in humans by Netherwood's group at Newcastle University (UK). The transgenic DNA was found in the small intestine of volunteers fed with GM soy. In three of seven cases there was low frequency gene transfer from GM soy to intestinal microflora.[66] Though the authors stress the importance of document-ing, for the first time in humans *in vivo*, survival of the transgenic construct and conservation of its biological activity that permitted effective transformation of saprophytic bacteria, they ventured that the phenomenon was unlikely to pose a health threat. Other commentators[67] disagreed, sustaining that the statement could be true in the specific case, for that particular gene construct (transgene *epsps*), but may not be true in the case of genes coding for antibiotic resistance. The surprising thing was the high frequency of transfer, three out of seven subjects, suggesting that the phenomenon is not as rare as we are led to think. It was objected that the study had intrinsic limitations, since it was conducted on patients with short intestines (ileostomized), whereas the transgene does not seem to survive its passage in the colon. This observation raises an even more worrying question: if intestinal length is what counts, may not the phenomenon affect children and adolescents, who have significantly shorter intestines than adults, in an even more significant way?

In confirmation of the correctness of these first results, we should recall that gene transfer has also been documented after eating GM microorganisms. *Lacto-bacillus*, widely used in processing human food (cheese, milk derivatives, yeasts) and feedstocks, and as a pharmaceutical (probiotic), is eaten live. The risk in the case of modified lactobacillus is that the bacteria transfer their gene constructs to intestinal microflora. These bacteria were not destroyed in the stomach and made up 2% of faeces of volunteers enrolled in a study by Brockmann and colleagues.[68] The marker gene was detected in faeces up to 4 days after administration of the microbes, when the latter were no longer alive. This is clear evidence that DNA even survives the death of the vector cells and can spread to microbe populations, exploit-ing mechanisms different from transformation, such as conjugation via plasmids

[65]S. M. Martin-Orue, A. G. O'Donnell, J. Arino, T. Netherwood, H. J. Gilbert & J. Mathers, 'Degradation of transgenic DNA from genetically modified soya and maize in human intestinal simulations', *Br. J. Nutr.* 87, 533–542 (2002).
[66]T. Netherwood, S. M. Martin-Orue, A. G. O'Donnell, S. Gokling, J. Graham, J. C. Mathers & H. J. Gilbert, 'Assessing the survival of transgenic plant DNA in the human gastrointestinal tract', *Nat. Biotechnol.* 22, 204–209 (2004).
[67]J. Heritage, 'The fate of transgenes in the human gut', *Nat. Biotechnol.* 22, 170–172 (2004).
[68]E. Brockmann, B. L. Jacobsen, C. Hertel, W. Ludwig & K. H. Schleifer, 'Monitoring of genetically modified *Lactococcus lactis* in gnotobiotic and conventional rats using antibiotic resistance markers and specific probe or primer based methods', *System Appl. Microbiol.* 19, 203–212 (1996).

and transposons.[69] Conjugational transfer of GM lactococci to saprophytic bacteria composing normal intestinal flora has been widely documented by studies *in vitro* and in gnotobiotic mice.[70,71]

Besides offering solid motivations for concern about the possibility of transfer of antibiotic resistance genes, this data raises new questions. The possibility of absorption of transgenic DNA and its subsequent incorporation in the host genome is far from being remote, since it has been confirmed for other types of foreign DNA, as we saw. Once inserted in the chromosome, the transgene can give rise to a series of genetic modifications, from silencing specific sequences and over-expression of others, to increased incidence of mutations. Together these modifications can produce unpredictable and harmful effects.[72]

Soy Tasting of Coconut

The seed company Pioneer Hi-Breed set out to "enrich" the nutritional value of soy by "improving" the genome through insertion of a Brazilian coconut gene. This gene would theoretically increase synthesis of albumin-2s, a protein particularly rich in the amino acids methionine and cysteine, the former is essential (it cannot be synthesized by the human body). Routine biochemical tests failed to detect any allergenic effects of the new product, despite the fact that this polypeptide is known to be a potent allergen that triggers a wide range of allergic reactions, ranging from itching to sudden death.[73] A subsequent test found that the allergenic power of the new product was even greater than that of coconut, though only a single gene had been transplanted into soy. Pioneer was forced to withdraw the new seed after publication of the data by the *New England Journal of Medicine*, losing millions of dollars that had been invested in the new variety. The company tried in vain to defend its product, but its arguments were weak. According to the multinational, this soy was for animals, but there is no way of

---

[69]T. Netherwood, R. Bowden, P. Harrison, A. J. O'Donnell, D. S. Parker & H. J. Gilbert, 'Gene transfer in the gastrointestinal tract', *Appl. Environ. Microbiol.* 65, 5139–5141 (1999).

[70]M. Gruzza, M. Fons, M. F. Ouriet, Y. Duval-Iflash & R. Ducluzeau, 'Study of gene transfer *in vitro* and in the digestive tract of gnotobiotic mice from *Lactobacillus lactis* strains to various strains belonging to human intestinal flora', *Microb. Releases* 2, 183–189 (1994).

[71]C. A. Alpert, D. D. Mater, M. C. Muller, M. F. Ouriet, Y. Duval-Iflash & G. Corthier, 'Worst-case scenarios for horizontal gene transfer from *Lactococcus lactis* carrying heterologous genes to *Enterococcus faecalis* in the digestive tract of gnotobiotic mice', *Environ. Biosafety Res.* 2, 173–180 (2003).

[72]K. Muller, H. Heller & W. Doerfler, 'Foreign DNA integration: genome-wide perturbations of methylation and transcription in the recipient genomes', *J. Biol. Chem.* 276, 14271–14278 (2001).

[73]D. N. Gillespie, M. D. S. Nakajima & M. D. Gleich, 'Detection of allergy to nuts by the radiallergo-sorbent tests', *J. Allergy Clin. Immunol.* 57, 302–309 (1976).

separating soy for humans from that for animals. It is also probable that the protein in question could be assimilated unaltered by animals and turn up on consumers' plates.

The episode is paradigmatic and prompts many interesting but worrying reflections. The first regards the alleged need to modify soy in order to increase its methionine and cysteine content. As pointed out by the *American Journal of Clinical Nutrition*:

> The amino acid profile of soy protein is unusually well-rounded for a plant protein. Comparison of a variety of soy proteins with ideal patterns published by the Food and Nutrition Board or the Food and Agriculture Organization demonstrates adequate quantities of the essential amino acids histidine, isoleucine, leucine, lysine, phenylalanine plus tyrosine, threonine, trypto- phan, and valine [...] nitrogen balance (N) of adults consuming textured soy protein, with and without methionine supplemen- tation, and ground beef (have been compared) [...] at higher intake of total protein (about $0.7\,g$ protein $\times kg^{-1} \times d^{-1}$, a level considered appropriate for adults), both soy products and beef resulted in essentially the same N balances.[74]

It is difficult to understand why a food that is already complete should be modified. Secondly, by FDA rules, allergy testing of the product was the producer's responsibility. These rules stipulate compulsory testing to ascertain the healthiness and harmlessness of the food, before it is marketed. The tests performed by the company were inadequate and insufficient. The product was marketed and even after publication of the report, nothing was done to indicate possible contraindications on the label. Only after a hard battle was the coconut soy withdrawn from the market. The ensuing debate about the need to label transgenic food involved consumers' associations, companies, researchers and the FDA. Consumers need to be able to identify and avoid potentially allergenic foods, but nothing concrete emerged to discipline a sector in which irregularities and confusion reign. Controls are few, inadequate, often superficial and completely in the hands of producers. The only reliable research is that conducted by university research groups, the few in this sector who are not bound in some way to agrochemical multinationals. The danger of modified soy was demonstrated, showing that antibodies developed by persons allergic to coconut interacted with proteins in the new soy beans. As the *New England Journal of Medicine* observed, the next case may occur under less ideal conditions and the public may be less fortunate. More information about incidence, prevalence, dietary exposure, antigenicity, immune response, diagnosis and treatment is needed before marketing of transgenic food can be authorized. This sensible advice is still largely ignored. Indeed, instead of learning from this

---

[74] J. W. Erdman Jr & E. J. Fordyce, 'Soy products and the human diet', *Am. J. Clin. Nutr.* 49, 725 (1989).

"unfortunate" episode (diluted versions of which are reported[75]), producers then had the brilliant idea of modifying soy by inserting the gene coding for beta-casein of cow's milk[76] and of transforming rice by transfecting it with a sequence from soy that synthesizes glycinin.[77] In both cases, the proteins are already known as allergens *per se*. It is impossible to understand the motivation for modifications that threaten health from the outset.

Urticaria and Other Pleasures

The first risk associated with the consumption of GM foods is a high incidence of allergy, as indicated in the above example.

The incidence of food allergies has increased significantly in the last 15 years.[78,79] In England, cases of food anaphylaxis increased from 6 to 41 cases per million between 1990 and 2000, and food allergies from 5 to 28 per million.[80] Among the many reasons are the profound changes in food production and processing, increased sensitivity to pollen that has caused cross-reactions with foods, contamination of foods with chemical residues, which while they may not be toxic in such concentrations may act as coadjuvants in causing allergic sensitivity, and finally, radical changes in eating habits with increasing consumption of exotic foods and foodstuffs foreign to culinary traditions.[81,82] In industrial countries, allergies are probably the most widespread chronic disease, affecting 15–30% of the population.[83] Allergic reactions to foods may be toxic or non-toxic. If the latter do

---

[75]S. L. Taylor, 'Food from genetically modified organisms and potential for food allergy', *Environ. Toxicol. Pharmacol.* 4, 121–126 (1997).

[76]P. J. Maughan, R. Philip, M. J. Cho, J. M. Widholm & L. O. Vodkin, 'Biolistic transformation, expression, and inheritance of bovine beta-casein in soy bean (*Glycine max*)', *In vitro Cell Dev. Biol.* 35, 344–349 (1999).

[77]T. Katsube, N. Kurisaka, M. Ogawa, N. Maryama, R. Ohtsuka, S. Utsumi & F. Takaiwa, 'Accumulation of soy bean glycinin and its assembly with the glutenins in rice', *Plant Physiol.* 120, 1063–1073 (1999).

[78]T. Keil, 'Epidemiology of food allergy: what's new? A critical appraisal of recent population-based studies', *Curr. Opin. Allergy Clin. Immunol.* 7, 259–263 (2007).

[79]D. A. Moneret-Vautrin, 'Modifications of allergenicity linked to food technologies', *Allerg. Immunol.* 30(1), 9–13 (1998).

[80]*Sempre più allergie*. Notiziario *Le Scienze*, 18.11.2003.

[81]D. A. Moneret-Vautrin, 'Les allèrgenes alimentaires et leurs modification par les technologies agroalimentaires', *Cah Agricultures* 6, 21–29 (1997).

[82]D. Jaffuel, P. Demoly, H. Dhivert-Donnadieu, J. Bousquet, F. B. Michel & P. Godard, 'Epidémiologie et genétique de l'asthme', *Rev. Mal. Resp.* 13, 455–465 (1996).

[83]The perception of allergy by patients and doctors is vastly different, since patients overestimate the incidence of the phenomenon. Often food intolerance is mistaken for allergy. Scientific studies give food allergy incidence figures of 2–3% in adults and about 8% in children and adolescents (cf. D. Jaffuel, P. Demoly & J. Bousquet, 'Les allergies alimentaires', *Rev. Fr. Allergol. Immunol.* 41, 169–186 (2001); I. Kimber & R. J. Dearman, 'Factors affecting the development of food allergy', *Proc. Nutr. Soc.* 61, 435–439 (2002)). However, the epidemiological data is underestimated, according to expert, because of the intrinsic difficulty of diagnosis.

not depend on the immune system, they are ascribed to metabolic defects, toxic reactions or deficiencies of intestinal enzymes and are considered intolerances;[84] when they are mediated by an immune reaction they are considered allergic reactions. Although these definitions are clear, they are often only theoretical: the clinical situation of individual cases is often different.

> The diagnosis of food allergy is still problematic, even in the case of atopy or IgE mediated hypersensitivity. There is a lack of standardized diagnostic procedures; the only test accepted as "gold standard" for confirmation of food allergy and in general for food intolerance, is a properly performed double blind placebo-controlled oral food challenge.[85]

Often even this is not sufficient and it is necessary to prolong the tests in time. Therapy is surprisingly simple (in theory): patients merely have to permanently avoid the foods in the case of allergy and avoid them for a period of time in the case of intolerance. The problem is particularly important in childhood, when a peak in the incidence of food allergies and intolerances is recorded, mostly related to the difficulty of effective therapy that avoids deficiencies and encounters patient compliance.[86]

The introduction of GM foods is not likely to improve this situation. Indeed, foods obtained by genetic recombination contain proteins and peptides, produced by the inserted or modified gene, which may objectively pose a threat.[87] The new proteins of bacterial, viral, animal and plant origin appear for the first time in the human diet and their allergenic potential is not only unknown, but also largely unpredictable on the basis of current tests. Genetic modification of foods can induce expression of allergenic substances in different ways and with different molecular mechanisms, many of which have already been well identified.

1. Many foods contain allergenic substances or immune irritants, albeit in low concentrations. At levels at which they are normally found in foods, these compounds produce minor effects that are usually negligible. Genetic manipulation of plants or animal cells can unexpectedly increase synthesis of this type of protein, bringing it to concentrations that can trigger an allergic reaction. If the gene in question synthesizes a protein already considered allergenic in the original food, foods transformed by insertion of genetic sequences can also

---

[84] V. Vigi & S. Fanaro, 'Le allergie alimentari nella prima infanzia', *Min. Ped.* 52(4), 215–225 (2000).

[85] C. Y. Pascual, 'Food allergy and intolerance in children and adolescents, an update', *Eur. J. Clin. Nutr.* 54(Suppl. 1), S75 (2000).

[86] P. B. Sullivan, 'Food allergy and food intolerance in childhood', *Ind. J. Pediatr.* 66(Suppl. 1), S37 (1999).

[87] C. Bindslev Jensen, 'Allergy risks of genetically engineered foods', *Allergy* 53, 58–61 (1998).

presumably become allergenic. Indeed, it is often difficult to exclude the possibility that a food causes an allergic reaction, whereas demonstrating the opposite calls for experiments on large population samples, which is neither ethical nor useful. When a food is suspected of causing allergy or this possibility cannot be excluded, it is wise and correct to inform consumers through a warning on the label. Such warnings would inevitably reduce sales, as the average consumer prefers non-modified products.

2. Many proteins expressed by genes inserted artificially in transgenic foods are not normally present in foods and have often never been part of the human diet. Thus their allergenic potential is unknown and cannot practically be evaluated by preclinical tests. As well, since nobody is likely to have been sensitized before by these new allergens, they do not induce a strong, immediately detectable reaction when first administered, only after repeated and prolonged consumption.

3. Genetic modifications can alter primary, secondary or tertiary structure of proteins, giving them new and unexpected properties. Indeed, even if a protein is not allergenic when it is produced in its original biological context, it can become so when the gene that encodes it is transposed artificially into a new organism. The primary structure (amino acid sequence) of long protein filaments is coded from DNA, but their folding, that ensures correct function through spatial form (secondary and tertiary structure), depends on their biological medium, specifically on what is known as the *morphogenetic field* of a cell. To induce a cell to produce foreign protein does not guarantee that the protein plays its normal role, nor that it does not acquire unexpected properties (toxic, allergenic …) due to incongruous spatial dispositions. This is specifically the case of proteins obtained through chimeric genes, i.e. genes obtained by fusing DNA segments from different species to form a new genomic unit. The allergenicity of these polypeptides cannot be deduced from the structural and biochemical characteristics of the parent proteins, since the chimeric protein can take completely unpredictable conformations different from those found in nature, two conditions that accentuate the foreign character of the molecule with respect to the living biological context and which make it even more probable that the protein be allergenic.

4. Different organisms do not necessarily have the same biochemical pathways of protein metabolism. A protein produced in a context different from the original may be broken down and processed incompletely, especially if it has anomalous configurations. This may lead to the formation of toxic metabolites or the release of fragments (haptens) that induce allergic reactions.

The situation is summarized by J.M. Wal, director of the Laboratoire d'Immuno-Allergie Alimentaire at Gif-sur-Ivette:

> No test such as the use of animal models, the analysis of structure, function and physico chemical properties is as yet available

to evaluate or predict the allergenicity of a "novel" protein in a wholly reliable and objective manner. No indication has yet suggested that novel foods, and particularly recombinant proteins or genetically modified foods, would be more (or less) allergenic than the corresponding conventional foods. No particular structure can be described as being solely and intrinsically allergenic. The predictive approaches to determining the allergenic potential of new foods should therefore be subject to case-by-case critical appraisal allied to mandatory implementation of monitoring of the potential postmarketing impact of these new foodstuffs on public health.[88]

It is no simple matter to correctly confirm an allergic reaction. As we saw, adverse reactions to foods can be non-immune (intolerances) or immune-mediated. The latter mostly depend on secretion of specific immunoglobulins (IgE) by type B lymphoid cells and manifest clinically minutes after ingestion of the allergenic food. The symptoms may include urticaria, angioedema, rhinoconjunctivitis, nausea, vomiting, diarrhoea and asthma. However, not all allergies are mediated by IgE: some are delayed, especially allergies to milk in children, and may induce a syndrome of enterocolitis with vomiting, diarrhoea and in adolescents, growth retardation.[89] The nosological picture is often complicated by chronic symptoms and is sustained by the formation of immune complexes and activation of lymphocytes. The pathogenetic mechanism is not yet clear, and the food components responsible have not been identified. Research into the allergenic potential of genetically engineered food has only concentrated on early reactions mediated by IgE,[90] which casts doubt on reassurances about the safety of GM foods. In actual fact, it is no easy matter to correctly diagnose food allergies. Available tests are of limited use and the most reliable parameter is still an accurate history that can reveal specific reactions to foods during the life of a person.[91] Suspected allergy can be confirmed by skin test or by assaying IgE in blood of patients by ELISA or radioimmunoassay. However, in some cases more complex examinations are needed, such as double-blind placebo-controlled food challenge (DBPCFC), in which samples of different foods are taken without the patient or doctor knowing their composition. The test is not routine because it requires expert personnel and special equipment in the case of anaphylactic reactions. Skin tests and assay of IgE are

[88] J. M. Wal, 'Assessment of allergic potential of novel foods', *Nahrung* 43(3), 168–171 (1999).

[89] A. M. Lake, Food protein-induced colitis and gastroenteropathy in infants and children, in: *Food Allergy: Adverse Reactions to Foods and Food Allergies*, D. D. Metcalfe, H. A. Sampson & R. A. Simon (Eds), Blackwell Scientific Publications, Oxford, UK (1997), pp. 277–286.

[90] D. D. Metcalfe, 'Introduction: what are the issues in addressing the allergenic potential of genetically modified foods?', *Environ. Health Perspect.* 111(8), 1110–1113 (2003).

[91] J. J. Jansen, A. F. Kardinaal, G. Huijberg, B. J. Vlieg-Boerstra, B. P. Martens & T. Ockhuizen, 'Prevalence of food allergy and intolerance in the adult Dutch population', *J. Allergy Clin. Immunol.* 93, 446–456 (1994).

unreliable: skin tests are often not sufficiently specific as the allergenic extracts are not reliably standardized[92] and the risk of systemic reactions is high. Their overall reliability is less than 50% due to a high frequency of false positives.[93]

Things are even more complex for potential sources of completely new allergens, such as GMOs. During a conference promoted by the US FDA, EPA and US Department of Agriculture in 1994, it was agreed that it was extremely likely that use of recombinant DNA in the development of new food varieties could generate unexpected new allergens.[94] No analytical methods of identifying *unknown* proteins exists and it is impossible to know reactive potential *a priori*. Current procedures can only compare new proteins (after lengthy separation and identification) with those known to be allergens, for similarities in structure, amino acid sequence, resistance to breakdown by enzymes or stomach acid, heat stability and molecular weight: parameters which cannot provide exhaustive information about the biological behaviour of the molecule, especially its capacity to elicit an immune reaction. This is why the strategies used for determining the allergenic potential of new foods, classified rationally in two consecutive protocols,[95] have been amply discussed and criticized.[96] Skin tests and DBPCFC on volunteers have aroused ethical problems; there is no agreement about the structural characteristics that define similarity between a known allergen and an unknown protein, since antigenicity may also be associated with altered fragments of the native protein, as in the case of milk peptides;[97] the criteria of heat stability and gastric digestion of proteins have been drastically limited,[98] since low resistance of a protein to breakdown by enzymes does not guarantee that it is innocuous: digested and denatured segments of polypeptides can elicit as strong an allergic reaction as the original protein.[99]

---

[92]R. Rhodius, K. Wickens, S. Cheng & J. Crane, 'A comparison of two skin test methodologies and allergens from two different manufacturers', *Ann. Allergy Asthma Immunol.* 88, 374–379 (2002).

[93]J. A. Bernstein, I. L. Bernstein, L. Bucchini, L. R. Goldman, R. G. Hamilton, S. Leher, C. Rubin & H. A. Sampson, 'Clinical and laboratory investigation of allergy to genetically modified foods', *Environ. Health Perspect.* 111(8), 1114–1121 (2003).

[94]Food and Drug Administration, Environmental Protection Agency, Department of Agriculture. *Conference on scientific issues related to potential allergenicity in transgenic food crops*, Annapolis, MD, April 18–19, 1994, Free State Reporting, Baltimore, MD (1994).

[95]D. D. Metcalfe, J. D. Astwood, R. Towsend, H. A. Sampson, S. L. Taylor & R. L. Fuchs, 'Assessment of the allergenic potential of foods derived from genetically engineered crop plants', *Crit. Rev. Food Sci. Nutr.* 36, S165–S186 (1996).

[96]J. M. Wal, 'Biotechnology and allergic risk', *Rev. Fr. Allergol. Immunol. Clin.* 41, 36–41 (2001).

[97]F. Maynard, R. Jost & J. M. Wal, 'Human IgE binding capacity of tryptic peptide from bovine α-lactoalbumin', *Int. Arch. Allergy Immunol.* 113, 478–488 (1997).

[98]T. J. Fu, 'Digestion stability as a criterion for protein allergenicity assessment', *Ann. N.Y. Acad. Sci.* 964, 99–110 (2002).

[99]I. Selo, L. Negroni, M. Yvon, G. Peltre & J. M. Wal, 'Allergy to bovine β-lactoglobulin: specificity of human IgE using CNDi derived peptides', *Int. Arch. Allergy Immunol.* 117, 20–28 (1998).

Paradoxically, fragmentation of a protein may increase its allergenic potential if it unmasks epitopes, previously hidden inside the original spatial conformation, making them accessible to IgE.[100]

What emerges from these studies is that allergenicity is not an intrinsic characteristic of molecules, since allergy seems to be a function of the nature, concentration and duration of exposure, rather than inherent allergenic power of the protein *per se*.[101] We only have imperfect ways of evaluating allergenic potential, the predictive value of which is unknown. The new version of the algorithm originally proposed considers these limitations and that the assessment of allergenicity cannot be taken for granted; to evaluate potential allergenicity it should be mandatory to perform specifically oriented experiments, case by case, on animals.[102,103] The inadequacy of current procedures is such that major allergy experts regard it as urgent that a series of measures be developed for further basic research and to correctly identify sources of risk.[104] This general uncertainty emerges particularly clearly when pleiotropic effects induced by transfected genes are considered, in other words effects on the functioning of other parts of the host genome that do not seem related to the modification sought. Indeed a modification may interfere with genes coding for endogenous allergens, increasing or modifying production of endogenous allergens of the transgenic organism with respect to its natural counterpart.[105] Finally, one should remember that anomalous immune reactions can also be triggered by non-protein components of GM organisms.

Immune reactions are normally triggered by substances recognized by the body as foreign, and involve many cell effectors (macrophages, T-helper and suppressor lymphocytes, natural killer cells, B lymphocytes) and sequential or concomitant expression of a multitude of gene sequences. Autoimmune reactions, often severe and invalidating, are those in which the immune reaction ends up being directed against certain antigens of the body, a sort of revolt against itself. The research group of Suzuki, an authority in this field, demonstrated that transfection involving

---

[100] P. Spuergin, M. Walter, E. Schiltz, K. Deichmann, J. Forster & H. Mueller, 'Allergenicity of α-caseins from cow, sheep, and goat', *Allergy* 52, 293–298 (1997).

[101] G. Bannon, T.-J. Fu, I. Kimber & D. M. Hinton, 'Protein digestibility and relevance to allergenicity', *Environ. Health Perspect.* 111(8), 1122–1124 (2003).

[102] V. E. Prescott & S. P. Hogan, 'Genetically modified plants and food hypersensitivity diseases: usage and implication of experimental models for risk assessment', *Pharmacol. Therap.* 111, 374 (2006).

[103] FAO/WHO, Evaluation of allergenicity of genetically modified foods. Report of a joint FAO/WHO Expert Consultation of Allergenicity of Foods Derived from Biotechnology, 22–25 January 2001, Rome, Italy, available at: http://www.fao.org/es/esn/gm/allergym.pdf.

[104] M.-J. K. Selgrade, I. Kimber, L. Goldman & D. R. Germolec, 'Assessment of allergenic potential of genetically modified foods: an agenda for future research', *Environ. Health Perspect.* 111(8), 1140–1141 (2003).

[105] A. W. Burks & R. L. Fuchs, 'Assessment of endogenous allergens in glyphosate-tolerant and commercial soybean varieties', *J. Allergy Clin. Immunol.* 96, 1008–1010 (1995).

insertion of small fragments of DNA and/or RNA in a cell, stimulates anomalous expression of genes coding for proteins of major histocompatibility complex (MHC) types I and II, activation of which is indispensable for autoimmune reactions, and of genes synthesizing antigenic proteins that migrate to the surface of the cell membrane. The effect obtained does not depend on the nature of the sequence introduced and may also be triggered by very small fragments, equal to or larger than 25 base pairs. The authors of the study rightly concluded that this phenomenon can contribute to the development of autoimmune phenomena when the DNA plasmid is introduced during gene therapy and can also be important when recombinant DNA is used for vaccinations and other purposes. If transfection of any DNA in cells can cause such accentuated alterations, it is surprising that these results have never been recorded and it is important that they be considered in future.[106]

In view of the complexity of the situation and the diversified risks associated with allergenic potential of GMOs, it would be wise not only to do the tests currently envisaged, but also rigorous clinical studies to determine whether a new product can induce allergic reaction in consumers after an appropriate period of experimentation (post-market surveillance).[107] Irrespective of whether or not this is ethical, GM food could not be directly marketed except after a first phase in which its distribution were limited to certain geographical areas, so that epidemiological data on its safety could be obtained after a period of time. This strategy of post-market surveillance would enable most new allergens introduced into the diet to be identified, though it would only minimize risks, not guarantee complete cover. Though favoured by all, such investigations still only exist on paper.[108] The organizational and methodological difficulties are great, but above all, it is almost impossible to relate a given product to an analysed event, since no labelling is required in countries where GMOs are most widely consumed.

Adoption of labelling laws is opposed by companies producing GM products, and is not even seriously considered by the institutions responsible for protecting human health. Unfortunately, the rules approved by the FDA[109] only consider the 10 most common food allergens. They ignore others, consumption of which is currently less widespread, but the health implications of which are nevertheless

[106]K. Suzuki, A. Mori, K. J. Ishii, J. Saito, D. S. Singher, D. M. Klinman, P. R. Krause & L. D. Kohn, 'Activation of target tissue immune recognition molecules by double-stranded polynucleotides', Proc. Natl Acad. Sci. U. S. A. 96, 2285–2290 (1999).

[107]H. A. Kuiper, G. A. Kleter, H. P. J. M. Noteborn & E. J. Kok, 'Assessment of food safety issues related to genetically modified foods', Plant J. 27(6), 503–528 (2001).

[108]J. J. Hlywka, J. E. Reid, J. C. Munro, 'The use of consumption data to assess exposure to biotechnology-derived foods and the feasibility of identifying effects on human health through post-market monitoring', Food Chem. Toxicol. 41, 1273–1282 (2003).

[109]'Maryanski discusses next wave in biotech, premarket issues', Food Chem. News January 23, 34–35 (1995).

no less serious. Consumer protection is completely inadequate for the latter allergens.[110]

The EU nominated an *ad hoc* commission on this type of risk. Its work concluded with the issue of Directive 2003/89 on 10 November 2003,[111] in which the need to label all foods with correct information about composition, avoiding deceptive wording and advertising was stipulated. The risk of an increase in food allergies and intolerances is particularly serious if one considers that persons allergic or intolerant to a substance cannot know whether the antigens have somehow been introduced into other products they decide to eat. As far as intolerances are concerned, since onset is often insidious with clinical manifestations and symptoms that may initially be difficult to diagnose, their incidence may increase in an uncontrolled manner.[112] Routine biochemical tests can only detect full-blown allergy in 2% of adults and 8% of children, whereas at least one American in four is estimated to suffer from food intolerance of some sort.[113] This hypersensitivity to food is likely to involve an increasing percentage of the population as new proteins enter the diet and as the percentage of foods containing additives and other substances, often of plant, viral or bacterial origin, increases.[114] Under such conditions, the allergenic potential of newly added proteins is uncertain, unpredictable and unverifiable.[115]

Only now are we timidly realizing the widespread nature and importance of these disorders. The causes of some have been discovered, while most remain unclear. There is still much work to be done, though every day new results confirm our fears, particularly as regards the problem of diet for children and adolescents. Transgenic products from soy and other foods are ingredients of many foods and drinks, especially those favoured by children and strongly advertised. For example, milk obtained from animals treated with synthetic bovine growth hormone (Bst) is added to nearly all packaged cakes. Synthetic Bst has an extra methionine with respect to natural Bst. This difference, apparently insignificant for many molecular biologists, is loaded with consequences. Thirty percent of persons coming into contact with Bst develop antibodies against the molecule,[116] indicating that the

[110]General Accounting Office, Food quality and safety: innovative strategies may be needed to regulate new food technologies, General Accounting Office, Washington, DC (July 1993).

[111]Directive 2003/89/EC of 10/11/2003, modifying 200/13/EC regarding indication of ingredients contained in food products.

[112]H. A. Sampson, 'Food allergy: Part I: Immunopathogenesis and clinical disorders', *J. Allergy Clin. Immunol.* 103, 717–728 (1999).

[113]H. A. Sampson & D. D. Metcalfe, 'Food allergies', *JAMA* 268, 2840 (1992).

[114]H. A. Sampson, L. Mendelson & J. P. Rosen, 'Fatal and near-fatal anaphylactic reactions to food in children and adolescents', *N. Engl. J. Med.* 327, 380 (1992).

[115]M. Nestle, 'Allergies to transgenic foods – question of policy', *N. Engl. J. Med.* 334(11), 726 (1996).

[116]*Physicians Desk Reference*, Baltimore, MD (1999).

"small" difference is sufficient to induce a reaction of allergic sensitization. By comparison, only 2% of persons ingesting natural Bst develop antibodies. This means that many consumers of milk and milk products containing Bst may develop intolerances or allergies that will be labelled "cause unknown". Despite this evidence, the FDA fails to see the need to properly study the allergenic potential of milk products and meat treated with Bst, or of foods derived from them. Also, marketing of these products was authorized without perceiving the need for labelling foods that could reasonably expose consumers to considerable risk.

The fact that many transgenic foods are marketed without rigorous labelling (required, however, in the EU) makes it almost impossible to assess their epidemiological impact. If GM components are less than 1% by weight, they are not indicated on the label and consumers cannot know that they are eating them. This happened for modified soy which is now mixed with natural soy and for which a massive increase in allergies (about 50%) was recorded only a year after its release. Habitual consumers of soy ended up with acne, eczema, digestive problems, chronic fatigue syndrome and lethargy.[117] The same is true of maize.

This instructive story concerns products of the company Aventis. In 2000, Aventis inadvertently released GM maize, StarLink™, originally limited to use in feedstocks because it was considered potentially allergenic. "Errors" of this kind are frequent. We saw a similar "accident" by Syngenta with Bt10 maize. StarLink maize contained a modified sequence (Cry9Ca1) of the Cry9C gene obtained from *Bacillus thuringiensis*. The product had insecticide properties that protected the modified maize plants. Studies to determine allergenic effects were initially negative, but it was soon realized that the testing method was far too simple, because the original gene produces a pro-toxin 129.8 kDa long that has to be cut in order to be activated, whereas the modified gene inserted in all cells of the plant directly synthesized the active toxin, which is only 68.7 kDa long and expresses its action irrespective of the metabolic capacity of the insect. Moreover, the tests were not conducted on the protein obtained from the modified maize (this was too complicated) but on one produced by bacteria. The allergy and toxicity tests were conducted on these samples. Only later was it realized that the protein Cry9Ca1 is glycosylated by maize (but not by bacteria)[118] and this discovery not only showed that the two polypeptides were not equivalent but also raised the doubt that the substance produced by maize could be allergenic. In fact, glycosylation is a structural characteristic of many known allergens: glycosyl groups mediate the reaction

---

[117] M. Towsend, 'Why soya is a hidden destroyer', *Daily Express*, 12 March (1999).
[118] B. Lambert, L. Busse, C. Decock, S. Jansen, C. Piens, B. Saey, J. Seurinck, C. Van Adenhove, J. Van Rie & S. Van Vliet, 'A *Bacillus thuringiensis* insecticidal crystal protein with a high activity against members of the family Noctuidae', *Appl. Environ. Microbiol.* 62, 80–86 (1996).

between antigen and antibody and ensure protein stability.[119] Indeed, the protein Cry9Ca1 expressed by StarLink maize was stable to heat and to peptic digestion.[120] In 1998, EPA only authorized its release for animal consumption for other reasons.[121] Nevertheless, the maize was marketed without the necessary warnings and was subsequently found in various products for human consumption. The cases of allergy and symptoms it caused are not known, but the FDA had to admit that in at least 28 cases of severe poisoning and gastrointestinal allergies, the cause was probably ingestion of this food.[122] The number was initially much larger, but a commission of the FDA and the Center for Disease Control of Atlanta selected only those in whom the allergic reaction was indisputable. A series of examinations were done in this group of subjects to see if they had developed IgE against the protein Cry9Ca1. The results were negative, but again the test was performed with a protein obtained from bacteria and not from the plant, that as we saw, produces a structurally different polypeptide. The Scientific Advisory Panel of EPA was forced to revise the data and concluded that the tests lacked the reliability necessary to conclude that the patients had not developed specific IgE antibodies against Cry9Ca1.[123] Allergic reactions can also occur by mechanisms that are not IgE-mediated. Thus in this case it is clear that the tests were insufficient[124] to explain the pathogenesis of the food allergies, which were, however, correctly recorded as such by FDA. As suggested, the allergenicity of transgenic proteins probably only emerges as a consequence of interactions with additives or other compounds found in food, since the food matrix significantly influences the breakdown of transgenic DNA and the possible formation of complexes with pathological potential.[125]

The random link between exposure and observed pathology was nevertheless considered sufficient by the FDA, which cancelled the authorization, ordered the

---

[119]R. Va Ree, M. Cabanes-Macheteau, J. Akkerdaas, J. P. Milazzo, C. Loutelier-Bourhis, C. Rayon, M. Villalba, S. Koppelman, R. Aalberse & R. Rodriguez, 'Beta(1,2)-xylose and alpha(1,3)-fucose residues have a strong combination in IgE binding to plant glycoallergens', J. Biol. Chem. 275, 11451–11458 (2000).
[120]L. Bucchini & L. R. Goldman, 'StarLink corn: a risk analysis', Environ. Health Perspect. 110, 5–13 (2002).
[121]US Environmental Protection Agency, 'Bacillus thuringiensis subspecies tolworthi Cry9c protein and the genetic material necessary for its production in corn; exemption for the requirement of a tolerance', Fed. Reg. 63, 28258–28261 (1998).
[122]Washington Post, 13 October 2000.
[123]US EPA FIFRA Scientific Advisory Panel, A set of scientific issues being considered by the Environmental Protection Agency regarding assessment of additional scientific information concerning StarLink [TM] corn. SAP Report No. 2001-09. US Environmental Protection Agency, Arlington, VA (2001).
[124]S. A. Sutton, A. H. Assa'ad, C. Steinmetz & M. E. Rothenberg, 'A negative double-blind placebo-controlled challenge to genetically modified corn', J. Allergy Clin. Immunol. 112, 1011–1012 (2003).
[125]V. Siruguri, B. Sesikeran & R. V. Bhat, 'StarLink genetically modified corn and allergenicity in an individual', J. Allergy Clin. Immunol. 113(5), 1003–1004 (2004).

withdrawal of StarLink and contaminated products from the market and prohibited future growing of the crop. The danger associated with GM maize also convinced the governor of Iowa and 16 others, who fined Aventis and ordered it to pay about a billion dollars in damages.[126]

The story of StarLink maize demonstrates many things: the inability of companies to monitor production and marketing of transgenic products, the inadequacy of government controls, the threat of allergy posed by proteins introduced for the first time in the human diet, the limits of diagnostic testing and the unreliability of post-market surveillance. Had it been considered that high exposure to Bt induces lethal lung infections in mice[127] and triggers allergic reactions in personnel involved in the production and application of pesticides containing Bt, more could probably have been done. At Cincinnati University, Bernstein demonstrated that exposure to spray containing proteins extracted from Bt caused allergic sensitization of the skin, with a dose- and time-dependent increase in specific IgE and IgG antibodies.[128] The response was directed prevalently against the protein Cry1A, not Cry9C expressed by modified maize. It is not known whether there is cross-reactivity between the two proteins, but it is probable, because structural homology is about 75%. Skin sensitivity to Bt protein could precede full IgE-mediated allergic reaction. Since various strains of *B. thuringiensis* have close molecular correlations[129] with other bacteria, such as *B. cereus*, a known food pathogen, and *B. anthracis*,[130] the agent of anthrax, it is conceivable that allergic sensitization may also extend to them and other microbes, such as *B. subtilis*, that would end up sharing cross-reactive allergenic epitopes and spreading the allergic phenomenon to innocuous components of intestinal

---

[126] "Tom Miller, Minister for Justice of Iowa inflicted defeat worth a billion dollars that represents the collapse of consumer faith in the GM food industry on the American market. Justice Miller, flanked by colleagues in 16 other states, condemned the company STARLINK, a branch of the French Aventis SA of Strasburg. Hundreds of federal inspections proved that modified maize marketed in 1999 and 2000 was toxic for domestic animals and could cause allergies in humans. The inspectors examined potato chips, cornflakes, syrups and even baby food. STARLINK could do nothing but accept the scientific results and pay the astronomical damages." Our translation from: La Stampa, 18 January 2001: *Usa, stop al grano transgenico. Multa record: "Fa male agli animali."* [USA stop to transgenic grain. Record fine: "It harms animals"].

[127] A. J. Trewavas, 'Much food, many problems', *Nature* 402, 231–232 (1999).

[128] I. L. Bernstein, J. A. Bernstein, M. Miller, S. Tierzieva, D. I. Bernstein, Z. Lummus, M.-J. K. Selgrade, D. L. Doerfler & V. L. Seligny, 'Immune response in farm workers after exposure to *Bacillus thuringiensis* pesticides', *Environ. Health Perspect.* 107(7), 575–582 (1999).

[129] P. H. Damgaard, H. D. Larsen, B. M. Hansen, J. Bresciani & K. Jorgensen, 'Enterotoxin-producing strains of *B. thuringiensis* isolated from food', *Lett. Appl. Microbiol.* 23, 146–150 (1996).

[130] A. J. Trewavas, 'Toxins and genetically modified food', *Nature* 355, 931 (2000).

flora.[131,132] Thus the possibility of previous exposure and sensitization to proteins, that could be cross-reactive with the new proteins contained in GM foods, should be considered a major risk factor before general release of these foods into the human food chain.[133]

These considerations are true for foods modified to express proteins with unknown or suspected allergenicity. They are even more true when GMOs express proteins not known to be allergenic. In this case, tests on animal models should be considered absolutely necessary, since for ethical reasons it is impossible to test the allergenic potential of proteins on humans.[134] Tests on animals entail ingestion of the food to be tested in the context of a balanced diet. Immune response is assessed by assaying IgE, reactions mediated by type II T-helper lymphocytes and clinical signs that best mimic those associated with allergic reactions in humans. These tests have obvious limitations, including the fact that negativity does not guarantee that an allergic reaction cannot occur in humans.[135] At the moment, together with other types of tests, it is the best approach to endeavour to define the allergenic profile of unknown proteins expressed by GM foods.[136] This recommendation was issued by the Codex Alimentarius Commission (Codex) that incorporated animal models in protocols already suggested.[137]

It is strange, however, that when these models indicate unequivocal evidence of allergenicity, they end up being disputed by those who proposed them, in attempts to sustain the concept that GMO are innocuous.

Pisellum sativum (pea) was recently modified by inserting a gene from *Phaseolus vulgaris* (bean) that expresses an inhibitor of the enzyme amylase (alpha-amylase-1) secreted by a vast range of parasites. Since this would make

[131] C. L. Johnson, I. L. Bernstein, J. L. Gallagher, P. F. Bonventre & S. M. Brooks, 'Familial hypersensitivity pneumonitis induced by *Bacillus subtilis*', Am. Rev. Resp. Dis. 122, 339–348 (1980).

[132] I. L. Bernstein, 'Enzyme allergy in populations exposed to long-term, low-level concentrations of household laundry products', J. Allergy Clin. Immunol. 49, 219–237 (1972).

[133] J. A. Bernstein, I. L. Bernstein, L. Bucchini, L. R. Goldman, R. G. Hamilton, S. Leher, C. Rubin & H. A. Sampson, op. cit.

[134] S. L. Taylor, 'Protein allergenicity assessment of foods produced through agricultural biotechnology', Annu. Rev. Pharmacol. Toxicol. 42, 99–112 (2002).

[135] V. E. Prescott & S. P. Hogan, 'Genetically modified plants and food hypersensitivity diseases: usage and implications of experimental models for risk assessment', Pharmacol. Therap. 111, 374–383 (2006).

[136] I. Kimber, R. J. Dearman, A. H. Penninks, L. M. J. Knippels, R. B. Buchanan, B. Hammerberg, H. A. Jackson & R. M. Helm, 'Assessment of protein allergenicity on the basis of immune reactivity: animal models', Environ. Health Perspect. 111(8), 1125–1130 (2003).

[137] R. E. Goodman, S. L. Hefle, S. L. Taylor & R. van Ree, 'Assessing genetically modified crops to minimize the risk of increased food allergy', Int. Arch. Allergy Immunol. 109, 136–142 (2002).

peas secrete a completely new protein (the inhibitor alphaA1), research was done to test the safety of the new GMO. Unfortunately it was discovered that transgenic expression of the bean protein in peas leads to synthesis of a modified form of alphaA1. This is not surprising because as we have seen, gene modifications may have unpredictable pleiotropic effects that cannot be attributed mechanistically to the gene modification. Alpha A1 synthesized by peas is not only different from that of the native plant, the bean, but has altered antigenic characteristics.[138] Mice fed with GM peas developed lung lesions due to hypersensitization and alphaA1-specific inflammation mediated by Th-2 lymphocytes and elevated levels of specific IgG antibodies that among other things had cross-immunoreactivity to other antigens in various other foods. Animals on diets supplemented with non-transgenic peas or beans (that produce the native inhibitor) did not show any kind of alteration. In the peas, the transgenic protein was subject to activation involving a series of post-translational modifications, including proteolytic cleavage (that removes certain amino acids from the original sequence) and above all glycosylation. The latter biochemical reaction consists of covalent bonding of one or more carbohydrate molecules to the protein skeleton. It gives the protein stability and resistance but unfortunately, it also often renders it antigenic. Other families of alpha-amylase inhibitors of cereals are known to elicit immune reactions with elevated levels of IgE.[139] Amylase inhibitors are anti-nutritional factors of cereals and from the medical point of view it would be better if humans did without them. Indeed, alpha A1 produced by transgenic peas has the same enzyme activity as that produced by beans, and the latter is known for inducing certain gastrointestinal dysfunctions.[140] Long cooking attenuates anti-amylase activity of the protein, eliminating the anti-nutritional property, but does not affect its allergenicity or its capacity to induce cell-mediated inflammation. Overall, these studies showed that transgenic expression of foreign proteins in plants can lead to the synthesis of structural variants with altered immunogenicity.[138] The study conducted on modified peas aroused much concern and triggered vehement reactions in politicians seeking easy fame. This occurred in Australia, where the research was conducted in universities and the Commonwealth Scientific and Industrial Research Organization (CSIRO). However, it is difficult to deny that the results raise serious doubts about the safety of the product in question. In this context, quibbling about the difference between

---

[138] V. E. Prescott, P. M. Campbell, A. Moore, J. Mattes, M. E. Rothemberg, P. S. Foster, T. J. V. Higgins & S. P. Hogan, 'Transgenic expression of bean α-amylase inhibitor in peas results in altered structure and immunogenicity', *J. Agric. Food Chem.* 53, 9023–9030 (2005).

[139] R. Sanchez-Monge, L. Gomez, D. Barber, C. Lopez-Otin, A. Armentia & G. Salcedo, 'Wheat and barley allergens associated with bakers asthma', *Biochem. J.* 281, 401–405 (1992).

[140] P. Layer, G. L. Carlson & E. P. DiMagno, 'Partially purified amylase inhibitor reduces starch digestion *in vitro* and inactivates intraduodenal amylase in humans', *Gastroenterology* 88, 1895–1902 (1985).

immunogenicity and allergenicity or casting doubt on the reliability of animal models, as published in an editorial of *Nature Biotechnology*,[141] is a despicable expedient for not tackling the basic questions raised by the discovery: transgenesis can induce unexpected modifications, loaded with consequences and unknowns that cannot be predicted or sufficiently identified by tests available today. And that is that.

This panorama reveals a picture that is anything but reassuring: there are no known or validated models or methods for assessing or predicting allergic risk produced directly or indirectly by genetic modifications:[142] methods developed for clinical purposes are now used to predict allergenicity or for post-market surveillance.[143] There are no reliable rules for ascertaining allergenicity:[144]

> The only way to confirm that a transgenic protein is or is not an allergen is to test it in large numbers of people. But of course, large-scale human testing isn't practical or ethically possible.[145]

It is simplistic to deny that protocols for determining risk and allergenic potential are inadequate.[146] As aptly pointed out by J.M. Wal, an authority on food allergies, our limited experience with GM foods makes it impossible to formulate solid hypotheses about increased or new allergic risk related to their consumption.[147]

It is deplorable that procedures of assessment of allergic risk were activated so late, only becoming systematic in 2002[148] after a long series of "accidents" and mobilization of public opinion and large sectors of the scientific community. This risk should have been assessed before GM foods were released for human

[141]'Genetically modified mush', *Nat. Biotechnol.* 24, 2 (2006).

[142]J. M. Wal, 'Biotechnology and allergic risk', *Rev. Fr. Allergol. Immunol. Clin.* 41, 36–41 (2001). Allergenic risk of foods or proteins also varies with geographic region. Rice, for example, does not provoke allergic reactions in Europe or the USA, but in Japan, where it is consumed daily, it poses problems. This sustains the opinion of many allergy specialists that allergic sensitization depends significantly on dose and the chronic nature of exposure to the potential allergen (cf. P. Celec, 'Biological and biomedical aspects of genetically modified food', *Biomed. Pharmacother.* 59, 531–540 (2005)).

[143]J. A. Bernstein, I. L. Bernstein, L. Bucchini, L. R. Goldman, R. G. Hamilton, S. Leher, C. Rubin & H. A. Sampson, *op. cit.*

[144]G. Lack, 'Clinical risk assessment of GM foods', *Toxicol. Lett.* 127, 337–340 (2002).

[145]C. W. Schmidt, 'Genetically modified foods: breeding uncertainty', *Environ. Health Perspect.* 113(8), A527–A533 (2005). The statement often heard, that "GMOs have been consumed by several hundred million persons for 3–4 years without any serious health problems" (B. E. B. Moseley, 'How to make foods safer – genetically modified foods', *Allergy* 56(Suppl. 67), 61–63 (2001)) is a completely unscientific oversimplification used for propaganda: where are the post-market studies supporting these conclusions?

[146]F. Sala, *Gli OGM sono Davvero Pericolosi?* Laterza, Bari (2005); G. Lack, *op. cit.*

[147]J. M. Wal, *op. cit.*

[148]M. Eubanks, 'Allergies à la carte. Is there a problem with genetically modified foods?', *Environ. Health Perspect.* 220(3), A130–A131 (2002).

consumption. Although the threat persists and will increase with per-head consumption and geographical spread, there is still time to pause for reflection and retrace our steps.

Poisonous Potatoes

An article in *The Lancet* (October 1999) by Pusztai's team at the Rowett Institute, Aberdeen University, Scotland, reported the consequences of prolonged consumption of potatoes transfected with the snowdrop gene:

> Diets containing genetically modified (GM) potatoes expressing the lectin *Galanthus nivalis* agglutinin (GNA) had variable effects on different parts of the rat gastrointestinal tract. Some effects, such as the proliferation of the gastric mucosa, were mainly due to the expression of the GNA transgene. However, other parts of the construct or the genetic transformation (or both) could also have contributed to the overall biological effects of the GNA-GM potatoes.[149]

The study showed specific lesions of intestinal villi and an effect that promoted growth of mucosal epithelium, providing the first experimental proof of biological damage related to consumption of GM food. The pathophysiological characteristics of the lesions provided a rational foundation that helped to explain gastrointestinal disorders and the onset of food intolerance syndromes. Pusztai's results were indirectly confirmed by studies originally directed at assessing possible side-effects of GM potatoes on insects. In research by the Scottish Crop Research Institute,[150] ladybirds, which are natural predators of plant parasites such as aphids, were fed for 12 days with aphids maintained on a diet of leaves of potatoes in which the snowdrop gene had been inserted to make them insect resistant. The ladybirds showed a sharp reduction in fertility, egg fertilization rate dropping from 95% to 78%, and their lives were shortened by 50%. Similar results were published by Fares a year earlier. Animals fed with GM potatoes transfected with the Bt gene to produce specific delta-endotoxin, showed a series of ileal enterocyte anomalies: cytoplasmic hyperplasia, nuclear hypertrophy, short microvilli and degeneration of mitochondria and endoplasmic reticulum.[151]

---

[149]S. W. Ewen & A. Pustzai, 'Effect of diets containing genetically modified potatoes expressing *Galanthus nivalis* lectin on rat small intestine', *The Lancet* 354, 1353–1354 (1999).

[150]N. A. E. Birch, I. E. Geoghegan, M. E. N. Majerus, J. W. McNicol, C. A. Hackett, A. M. R. Gatehouse & J. A. Gatehouse, 'Tri-trophic interactions involving pest aphids, predatory 2-spot ladybirds and transgenic potatoes expressing snowdrop lectin for aphid resistance', *Mol. Breed.* 1, 75–83 (1999).

[151]N. H. Fares & A. K. El-Sayed, 'Fine structural changes in the ileum of mice fed on delta-endotoxin-treated potatoes and transgenic potatoes', *Nat. Toxins* 6, 219–233 (1998).

Announcement of the results, that Pustzai ingenuously illustrated in a TV pro-gramme, produced a cascade of devastating effects. His research was blocked and he was immediately removed from the research institute. In the meantime, a fierce controversy broke out in the specialized press and mass media,[152] dividing the sci-entific community into those who denigrated the research in order to detract from the conclusions and those who supported the scientist and promoted a moratorium to block marketing of GMOs until their harmlessness was proven.[153] Paradoxically, the ferocity with which critics of Pustzai attacked his work backfired against propo-nents of GMOs. The research group at the Rowett Institute certainly did not intend to demonstrate that any GMO was harmful. Their message was simply to stress the need for proper testing before a food was marketed,[154] a reasonable request. Among the vehement reactions elicited, it was revealed that the Royal Society[155] condemned the research without even reading it! The episode demonstrated the hypocrisy of the biotech lobby; always ready to defend the right to "freedom of research" against any hint of imposition of ethical criteria or environmental pru-dence, and simultaneously ready to deny freedom of research to anyone whose research or opinions cast doubt on the truth and reliability of the new biotech religion.

However, negative effects of consumption of food containing GM plants have been observed in animal models and are beginning to appear in specialized journals, albeit timidly and with difficulty.

Manuela Malatesta, an enterprising young researcher at Urbino University, Italy, conducted a series of studies into the pathophysiological consequences of a diet including soy modified to impart resistance to glyphosate (Roundup Ready™ soy). Rats fed with the GMO for prolonged periods (up to 8 months) showed minor pancreatic alterations[156] (with reduction of pancreatic enzyme efficiency,[157] RNA activity and post-transcriptional RNA processing) and significant changes in liver cell nuclear ultrastructure (morphological modifications of nuclear pores, nucleolus and the nuclear membrane), indicative of increased metabolism and the capacity of a diet based on GMOs to specifically interfere with liver cell nuclear

---

[152] Compare the debate conducted in *The Lancet* (*The Lancet*, 1999, vol. no. 354, p. 1354, pp. 1314–1315, pp. 1725–1729 and p. 684).

[153] B. Christie, 'Scientists call for moratorium on genetically modified foods', *Br. Med. J.* 318, 483 (1999).

[154] J. M. Rhodes, 'Genetically modified foods and the Pustzai affair', *Br. Med. J.* 318, 1284 (1999).

[155] *The Lancet* 354, 684 (1999).

[156] M. Malatesta, M. Biggiogera, E. Manuali, M. B. L. Rocchi, B. Baldelli & G. Gazzanelli, 'Fine structural analyses of pancreatic acinar cell nuclei from mice fed on genetically modified soybean', *Eur. J. Histochem.* 47(4), 385–388 (2003).

[157] M. Malatesta, C. Caporaloni, L. Rossi, S. Battistelli, M. B. L. Rocchi, F. Tonucci & G. Gazzanelli, 'Ultrastructural analysis of pancreatic acinar cells from mice fed on genetically modified soybean', *J. Anat.* 201, 409–415 (2002).

activity.[158] Similar results were also found in rat testicles.[159] The alterations were reversible,[160] but still demonstrate how brief exposure to GM food can cause major ultrastructural modifications in cells of the adult organism. It is not clear whether these results should be ascribed to the direct action of the transgene, possible residues of glyphosate or incompletely specified pleiotropic effects related to the reduced expression of phytooestrogens documented in GM soy.[161,162] A previous study by Monsanto[163] was unable to detect any alteration in rats fed with wheat modified with the same transgene as Roundup ready soy: in that case the rats were fed the product for too brief a period (only 13 weeks), a limitation that also seems to apply to other animal studies that failed to reveal any negative effect.[164] However, the results obtained by the young Italian researcher clearly show that consumption of GM foods can cause significant, measurable, structural modifications, though the pathogenic potential of prolonged exposure is not known. Studies into this aspect were planned by Malatesta, but her funds were cut and the research hindered in an unimaginable variety of ways. Malatesta gave in and transferred to another university. Will we hear the end of this story?

GM maize is likewise not without its doubts. A recent study by different Italian research groups found that sheep fed with Bt maize did not show macroscopic alterations of organs and tissues but a wide range of unexpected modifications. Biochemical analysis of rumen epithelium showed signs of proliferative activation with increase in Ki67 titres, while liver and pancreas cell nuclei contained an elevated number of heterochromatin granules, indicating perturbation of gene translation mechanisms. The meaning and implications of these results are unclear. However, the results produced by G.-E. Séralini, a pioneer of GMO studies, in

[158] M. Malatesta, C. Caporaloni, S. Gavaudan, M. B. L. Rocchi, S. Serafini, C. Tiberi & G. Gazzanelli, 'Ultrastructural morphometric and immunocytochemical analyses of hepatocyte nuclei from mice fed on genetically modified soybean', *Cell Struct. Function* 27, 173–180 (2002).

[159] L. Vecchio, B. Cisterna, M. Malatesta, T. E. Martin & M. Biggiogera, 'Ultrastructural analysis of testes from mice fed on genetically modified soybean', *Eur. J. Histochem.* 48(4), 449–454 (2004).

[160] M. Malatesta, B. Baldelli, S. Battistelli, C. Tiberi, E. Manuali & M. Biggiogera, 'Reversibility of hepatocyte nuclear modifications in mice fed on genetically modified soybean', *Eur. J. Histochem.* 49(3), 237–241 (2004).

[161] M. A. Lappé, E. B. Bailey, C. Childress & K. D. R. Setchell, 'Alterations in clinically important phytooestrogens in genetically modified herbicide-tolerant soybeans', *J. Med. Food* 1, 241–245 (1999).

[162] A. Cassidy, 'Potential risks and benefits of phytoestrogen-rich diets', *Int. J. Vitamin Nutr. Res.* 73, 120–126 (2003).

[163] B. Hammond, R. Dudek, L. Lemen & M. Nemeth, 'Results of a 13 week safety assurance study with rats fed grain from glyphosate tolerant corn', *Food Chem. Toxicol.* 42, 1003–1014 (2004).

[164] W. Hashimoto, K. Momma, H.-J. Yoon, S. Ozawa, Y. Ohkawa, T. Ishige, M. Kito, S. Utsumi & K. Murata, 'Safety assessment of transgenic potatoes with soybean glycinin by feeding studies in rats', *Biosci. Biotechnol. Biochem.* 63, 1942–1946 (1999).

rats fed with MON863 maize were even more striking. This variety of GM maize was the subject of fierce controversy, culminating in the sentence of the Munster Court of Appeal, Germany, ordering access to the results of testing conducted by Monsanto on experimental animals, on which the authorization conceded in 2005 was based. The results were published by the research group of Hammond,[165] supervised by Monsanto, and were analysed again by Séralini after winning a further legal action that Monsanto fought to the very end. The statistical method used by Hammond *et al.* did not bring out significant differences and was widely and justifiably criticized, whereas Séralini's approach produced several unpleasant surprises, though in this case they were not unexpected. The rats fed with MON863 showed clear signs of liver and kidney toxicity (18 cases out of 20, not considered "relevant" by the Monsanto experts!), increased lipidemia (triglycerides up to 24–40%), reduced excretion of sodium and phosphorus and retarded weight increase. It emerged clearly that

> [...] It appears that the statistical methods used by Monsanto were not detailed enough to see disruptions in biochemical parameters, in order to evidence possible signs of pathology within only 14 weeks. Moreover, the experimental design could have been performed more efficiently to study subchronic toxicity, in particular with more rats given GMOs in comparison to other groups. Considering that the human and animal populations could be exposed at comparable levels to this kind of food or feed that has been authorized in several countries, and that these are the best mammalian toxicity tests available, we strongly recommend a new assessment and longer exposure of mammals to these diets, with cautious clinical observations, before concluding that MON863 is safe to eat.[166]

The competent authorities did not follow this advice. Furthermore, Séralini and his laboratory recently provided new insights and a compelling confirmation of these results. The study clearly revealed new side-effects linked to GM maize consumption for three GM strains of maize. The effects were sex- and often dose-dependent, and were mostly associated with the kidneys and liver, the diet detoxifying organs. Other effects were also noticed in the heart, adrenal glands, spleen and haematopoietic system, while other unintended direct or indirect metabolic consequences of the genetic modification were not excluded.[167]

---

[165]B. Hammond, J. Lemen, R. Dudek, D. Ward, C. Jiang, M. Nemeth & J. Burns, 'Results of a 90-day safety assurance study with rats fed grain from corn rootworm-protected corn', *Food Chem. Toxicol.* 44, 147–160 (2006).

[166]G.-E. Séralini, D. Collier & J. Spiroux de Vendômois, 'New analysis of a rat feeding study with genetically modified maize reveals signs of hepatorenal toxicity', *Arch. Environ. Contam. Toxicol.* 52(4), 596–602 (2007).

[167]J. S. de Vendômois, F. Roullier, D. Cellier & G.-E. Séralini, 'A comparison of the effects of three GM corn varieties on mammalian health', *Int. J. Biol. Sci.* 5(7), 706–726 (2009).

Similarly, scientific controversy also exists in relation to the safety of GM soybeans. While it has been reported that 356043[168] and 305423[169] soybeans are as safe as conventional non-GM soybeans, some authors are still concerned about the safety of GM soybeans and recommend that long-term consequences of GM diets and potential synergistic effects with other products and/or conditions be investigated.[170,171] A recent review by Dona and Arvanitoyannis is especially critical.[172] They remark that the results of most studies with GM foods indicate that they may cause common toxic effects (hepatic, pancreatic, renal and reproductive) and may alter blood, biochemical and immunological parameters. These authors also conclude that the use of recombinant GH or its expression in animals should be re-examined since it is demonstrated to increase IGF-1 which may in turn promote cancer. A harsh and unconvincing response to the review was recently published in the same journal.[173] This is only an example of the controversy, which remains unsettled at all levels.

Genomic Instability and Pleiotropic Effects

Gene transfer techniques of today do not permit insertion of a foreign DNA fragment to be guided to a precise position in the genome. The gene may end up in a blind region where it cannot express any function, or in sectors where its activity is influenced to different degrees by neighbouring genes. The spatial context in which the fragment is inserted therefore affects its expression. This is what is meant by "position effect". DNA is not a simple succession of segments of specialized nucleic acids associated in a random way, just as a city is not a succession of houses, shops and monuments: each occupies a definite, unique and essential place. It is true that buildings or even whole areas of cities could be moved without changing their physiognomy, but if we think of moving the Coliseum from the forum region of Rome to the periphery at *Centocelle*, anyone would agree that this would violate

---

[168]Y. Sakamoto, Y. Tada, N. Fukumori, K. Tayama, H. Ando, H. Takahashi *et al.*, 'A 52-week feeding study of genetically modified soybeans in F344 rats', *J. Food Hyg. Soc. Jpn* 48, 41–50 (2007) (in Japanese).

[169]B. Delaney, L. M. Appenzeller, S. M. Munley, D. Hoban, G. P. Sykes, L. A. Malley *et al.*, 'Subchronic feeding study of high oleic acid soybeans (event DP-3Ø5423-1) in Sprague–Dawley rats', *Food Chem. Toxicol.* 46, 3808–3817 (2008).

[170]J. A. Magaña-Gómez & A. M. de la Barca, 'Risk assessment of genetically modified crops for nutrition and health', *Nutr. Rev.* 67, 1–16 (2009).

[171]B. Cisterna, F. Flach, L. Vecchio, S. M. L. Barabino, S. Battistelli, T. E. Martin *et al.*, 'Can a genetically modified organism-containing diet influence embryo development? A preliminary study on pre-implantation mouse embryos', *Eur. J. Histochem.* 52, 263–267 (2008).

[172]A. Dona & I. S. Arvanitoyannis, 'Health risks of genetically modified foods', *Crit. Rev. Food Sci. Nutr.* 49, 164–175 (2009).

[173]C. Rickard, 'Response to "Health risks of genetically modified foods" ', *Crit. Rev. Food Sci. Nutr.* 50, 85–91 (2010).

the city's identity and have dramatic functional consequences. The same is true for chromosomes and DNA: the position occupied by a gene segment is not arbitrary. This does not mean that its position is obligatory, only that there are presumably positions incompatible with expression of the gene or others that deprive it of the necessary connections with regulator and operator genes. Undoubtedly, any living cell in which a new gene segment is inserted artificially makes the genome less stable and more susceptible to spontaneous mutations. Foreign genes inserted randomly cause DNA rearrangements and may alter the expression of other genes, leading to expression of slightly modified proteins, which for this very reason are probably toxic. Such modifications are presumably the source of the unexpected toxicity phenomena manifested by modified foods such as canola and sugar beet:[174] even a slight modification can have severe consequences, especially if it inadvertently reactivates blocked toxins, such as the glucosinolates and erucic acid of canola.

Dangers of this type do not seem to have been averted by new technologies designed to do just that. For example, chimeraplasty makes it possible to replace a single base in an exact position in a specific gene of a cell. A 25-base probe is constructed, homologous to a corresponding sequence of the gene to be modified, except for the base to be replaced. Once in the cell, the probe flanks the homologous DNA regions and a series of enzymes are used to make it replace the natural sequence. This technique has been used to modify a gene of tobacco, to make the tobacco plant resistant to herbicides. A codon which identified proline (CCA) in amino acid position 196 of the protein expressed by the gene was replaced by glutamine (CAA) and leucine (CTA). The results indicated successful substitution, but not in the sense hoped by the researchers. Instead of CTA (leucine) a TCA codon specifying serine was obtained, and instead of CAA (glutamine), ACA (threonine) was found. This is like randomly replacing words in a conversation: blade instead of bald, café instead of face, pleat instead of plate, shape instead of phase, and so forth. Babel! It is exactly what happened to the poor cell when it began to synthesize "phrases": the proteins were senseless or had a different or opposite meaning from that originally intended. According to the rule that only confusion can arise from confusion, the genomic heritage of the cell began to undergo spontaneous mutations at a frequency about 20 times that recorded in the control culture.[175]

Insertion of the transgene can determine pleiotropic effects. These are not necessarily toxic: they can be neutral or even desirable. What is puzzling is that these modifications occur unpredictably, causing significant fluctuations in the concentrations of different substances by mechanisms that nobody has yet succeeded in

---

[174] V. Tardieu, 'La génétique menace-t-elle l'alimentation?', *Dossier Science et Vie* 950, 33–35 (1996).
[175] P. R. Beetham *et al.*, 'A tool for functional plant genomics: chimeric DNA–RNA oligonucleotides cause *in vivo* gene-specific mutations', *Proc. Natl Acad. Sci. U. S. A.* 96, 8774 (1999).

identifying. The genetic modification of potatoes, for example, leads to a reduction in steroid glycosidic alkaloids. However, the effect is not due to the inserted transgene, nor to orientation of the coding sequence or the site of insertion. All explanations are only suppositions.[176]

The Unpredictable Is Always Unexpected

Tryptophan is an essential amino acid, meaning that dietary intake is necessary to avoid deficiency, which typically manifests as pellagra, a complex syndrome leading to death which for decades plagued rural southern Italy. Immediately after the unification of Italy, when the south was annexed to Piedmont, peasants and southern agriculture suffered great poverty. The etiopathogenesis of pellagra depends on deficiency of nicotinamide, a vitamin cofactor, produced from tryptophan. Usually this amino acid is well represented in proteins, however the peasants and labourers of the south lived almost exclusively on polenta, the main protein (zein) of which practically without tryptophan (Trp). Hard-won experience in the therapy of this scourge and of other vitamin deficiencies induced health authorities in industrialized countries to promote dietary supplements of vitamins and essential amino acids, particularly after World War II. In the meantime it was discovered that Trp was an indispensable precursor of two neurohormones, serotonin and melatonin, that control humour and the rhythm of sleeping and waking, respectively. In the first half of the 1980s, hydroxytryptophan, a drug consisting of an intermediate in the metabolic pathway between Trp, serotonin and melatonin, was marketed. Immediately afterwards, reports relating consumption of Trp and a new disease called eosinophilic–myalgia syndrome began to appear. The FDA intervened immediately (on this occasion), exhorting consumers to reduce consumption of food supplements and drugs based on Trp.[177] In the meantime, a first balance of the damage showed that the new syndrome had reaped 36 victims and caused innumerable invalidities and permanent paralyses: a thousand, two thousand, ten thousand cases according to different sources.[178] Old hypotheses that ascribed a role to Trp in the aetiology of bladder cancer were also revived and confirmed.[179] Today we know that the problem was due to a lot of Trp produced by Showa Denka company exploiting

---

[176]M. Stobiecki, I. Matysiak-Kata, R. Franski, J. Skala & J. Szopa, 'Monitoring changes in anthocyanin and steroid alkaloid glycoside content in lines of transgenic potato plants using liquid chromatography/mass-spectrometry', *Phytochemistry* 62, 959–969 (2003).

[177]A. N. Mayeno & G. J. Gleich, 'Eosinophilia-myalgia syndrome and tryptophan production: a cautionary tale', *Trends Biotechnol.* 12(9), 346–352 (1994).

[178]F. P. D'Arcy, 'L-Tryptophan: eosinophilia-myalgia syndrome', *Adverse Drug React. Toxicol. Rev.* 14, 37 (1995).

[179]M. Bizzarri, A. Catizone, M. Pompei, L. Chiappini & A. Laganà, 'Determination of urinary Trp and its metabolites along the nicotinic acid pathway by HPLC with UV detection', *Biomed. Chromat.* 4(1), 24–28 (1990).

a GM strain of *Bacillus amytoliquefaciens* to produce Trp on industrial scale.[180] The modification consisted in insertion of certain gene sequences that permit the microbe to express high quantities of Trp. This artifice clearly optimized the yield of the amino acid, which previously had to be extracted from a mixture of bacterial proteins. The FDA authorized sale of Trp in 1988. Conviction that the production process could not in any way affect the safety of the product was so ingrained that no preliminary testing was required: after all, Trp is innocuous and had been on the market for years, produced by traditional methods. Of course no labelling was prescribed to inform consumers or doctors that the product was obtained by new gene recombination techniques. The resulting disaster arrived like a bolt out of the blue. The problem depended on the nature of the genetic modification that enabled bacteria to produce extraordinary quantities of the amino acid in question. When the concentration of Trp exceeds certain intracellular levels, Trp molecules or their precursors tend to interact spontaneously among themselves, giving rise to extremely toxic dimers that are difficult to detect analytically. The dimer of Trp was found in lots produced by Denko in apparently negligible concentrations – less than 0.1% by weight – but sufficient to trigger a lethal syndrome in the human body.[181] Since the concentration of the contaminant was low, Trp produced by the new method was considered "substantially equivalent", *when it had clearly become lethal.* Scientific investigations did not reach an unequivocal conclusion about this much discussed episode. It is not known whether the pathology was due to the purification procedures, the presence of contaminants unrelated to the transgenic method used, or something else. The company hurriedly destroyed the incriminated bacteria and compensated all victims who took legal action.[182]

This instructive episode shows the limits of the safety measures taken, often with bureaucratic superficiality close to criminal negligence. Secondly, it reveals a problem that is difficult to solve: toxic "subproducts" elaborated by the metabolism of GMOs may not be detected in the final product approved for consumption. Analytical chemistry only enables what is sought, or what is presumed could be

---

[180]E. M. Kilbourne, R. M. Philen, M. L. Kamb & H. Falk, 'Tryptophan produced by Showa Denko and epidemic eosinophilia-myalgia syndrome', *J. Rheumatol. Suppl.* 46, 81–88 (1996).

[181]Showa Denko did not merely modify the production process of Trp by GM bacteria but streamlined and simplified purification of the amino acid obtained in culture. The incriminated bacterium was destroyed as soon as the connection was realized between recombinant Trp and eosinophilic myalgia syndrome, so it is not possible to know today whether contamination of Trp by its dimer was a direct consequence of the genetic manipulations or a result of changes in the purification procedure. Since the latter has also been used by companies whose Trp, completely devoid of toxic contaminants, was produced without recourse to bioengineering, it is more than likely that dimer production can be ascribed to genetic manipulation.

[182]D. Schubert, 'A different perspective on GM food (reply)', *Nat. Biotechnol.* 20, 1197 (2002).

present, to be measured, albeit with great precision. By definition, substances the existence of which is not contemplated cannot be detected until they have manifested. This fact alone invalidates the principle of "substantial equivalence" that companies appealed to in order to prevent labelling of transgenic food. The criterion of substantial equivalence is scientifically inadequate and only has sense at the level of TV talk shows: on the basis of what parameters is equivalence established? A food is initially considered equivalent to another only when it has a similar composition of fats, carbohydrates and proteins. Nothing is said about vitamins, trace elements and enzyme cofactors. With regard to the presence of toxic factors, only the best known are monitored: phytates, glucosilonates (in canola), solanin, tomatins, lectins, oxalates (in tomatoes and potatoes), protease inhibitors, isoflavones, phytates and lectins (soy), glycosidic alkaloids.[169] Unfortunately such data is rarely reported for GM products: in three out of four cases, no information is given;[183] besides, below certain concentrations or for unusual and unsuspected components, the principle of equivalence is of no help. It is worth recalling that even the meat of animals fed with meat and bone meal infected with mad cow disease was passed as "substantially equivalent" – at least until the prion responsible for the disease was looked for and found. How many other prions can we expect in the next few years?

Other Confirmations

Phenomena of this kind should not surprise us, since it is well known that genetic manipulation can induce major unexpected changes in metabolic processes and induce synthesis of new, potentially allergenic or toxic proteins. For example, several years ago, in order to accelerate the breakdown of carbohydrates, the genome of a yeast that had initially given promising results was modified. Under physiological cell conditions, an enzyme process is minutely regulated by many negative and positive feedbacks. Enzymes are generally proteins that rely on one or more cofactors, including vitamins. Enzyme efficiency depends on physical (temperature) and above all biochemical parameters (substrate availability, quantity of product, inhibitors, and so forth). Any variation in enzyme activity not balanced by these factors inevitably has repercussions on a cascade of processes started by the reaction catalysed by the enzyme. Excessively fast enzymic breakdown of sugars to pyruvate, an intermediate compound in the biochemical pathway from sugar to energy, not compensated by equally fast activation of a second enzyme cascade known as the respiratory chain, causes accumulation of pyruvate that is then transformed into lactate. Lactate is toxic for muscle fibre cells and needs oxygen to be metabolized. This is what happens during muscle exertion: work makes the

---

[183] W. K. Novak & A. G. Haslberger, 'Substantial equivalence of antinutrients and inherent plant toxins in genetically modified novel foods', *Food Chem. Toxicol.* 38(6), 473–484 (2000).

body burn sugars to produce ATP, the intercellular exchange currency. Carbohydrate breakdown is immediate but incomplete, with accumulation of pyruvate and lactic acid that causes a sensation of muscle fatigue. Breathing frequency increases and the body has to rest to get rid of excess lactate. Obviously this occurs less in athletes because exercise trains the biochemical machinery of muscles and causes a compensating increase in respiratory chain enzyme concentrations. As in muscle, the increased concentration of the yeast enzyme caused accumulation of the product of the reaction catalysed by the enzyme. The compound in question was an intermediate in the sugar breakdown process, which does not accumulate under physiological conditions because it is produced much more slowly than it is disposed of along the enzyme pathway. Unfortunately the product in question was methylglyoxal, a notorious carcinogen, and its concentrations were 30 times higher than those found in non-modified strains of yeast. Obviously the experiment was suspended and the idea of playing with such complex biochemical paraphernalia was given up, even if yeast is one of the simplest cells we know. Again the concept of "substantial equivalence" was radically contested by indisputable data. In their conclusions, the authors of the study demonstrated that they fully understood the portent of their experiments and the concern their results had created:

> In genetically engineered yeast cells, the metabolism is significantly disturbed by the introduced genes or their gene products and the disturbance brings about the accumulation of the unwanted toxic compound MG in cells. Such accumulation of highly reactive MG may cause a damage in DNA, thus suggesting that the scientific concept of "substantially equivalent" for the safety assessment of genetically engineered food is not always applied to genetically engineered microbes [...].Thus, the results presented may raise some questions regarding the safety and acceptability of genetically engineered food, and give some credence to the many consumers who are not yet prepared to accept food produced using gene engineering techniques.[184]

The scientific literature on GMOs has lately begun to raise the problem of unintended effects, those that have no apparent or immediate relation to the modifications made, since they depend more on perturbation of the system as a whole than on involvement of a single compartment or metabolic pathway. Some of these "anomalies" – which may often have major repercussions on health – have been investigated, producing a mass of information and many unexpected surprises.[185]

---

[184]T. Inose & K. Murata, 'Enhanced accumulation of toxic compound in yeast having glycolytic activity: a case-study on the safety of genetically engineered yeast', *Int. J. Food Sci. Technol.* 30, 141–145 (1995).
[185]V. Ahuja, M. Quatchadze, V. Ahuja, D. Stelter, A. Albrecht & R. Stahlmann, 'Evaluation of biotechnology-derived novel proteins for the risk of food-allergic potential: advances in the development of animal models and future challenges', *Arch. Toxicol.* 84, 909–917 (2010).

In canola modified genetically to express higher values of phytoene synthetase, many unexpected metabolic changes have been observed in tocopherol, chlorophyll, phytoene, fatty acids and above all in glycoalcohols (up 50%).[186] Soy modified to ensure resistance of the plant to glyphosate showed two anomalies: more accentuated synthesis of lignin (an increase in lignin makes soy less digestible and lowers its quality as food) at soil temperatures around 20°C, whereas at high soil temperatures (40°C) there is a dramatic drop in yield (down 40%).[187] Rice manipulated to produce provitamin A (Golden Rice) showed an exceptional increase in unexpected molecules (lutein, beta-carotene, zeaxanthin) during testing.[188] Modified wheat showed accentuated alteration of phospholipid metabolism due to increased expression of phosphatidylserine synthetase.[189] Cotton and tobacco modified to resist infections developed necrotic lesions due to an unintended increase in glucose oxidase,[190] a phytotoxic enzyme. GM potatoes showed different anomalies in tuber architecture and carbohydrate transport.[191] Contrasting results were also observed in glycoside alkaloid production in GM potatoes, with significant increases or decreases that presumably depended on different experimental conditions.[192,193]

Special concern has recently been expressed about endocrine interference by GMOs. Indeed, GM food associated with xenobiotics, such as pesticide residues and xenoproteins, could be harmful in the long term. The "low-dose hypothesis", accumulation and biotransformation of pesticides associated with GM food and the multiplied toxicity of pesticides and formulation adjuvants support this

---

[186]C. K. Shewmaker, J. A. Sheeley, M. Daley, S. Colburn & D. Y. Ke, 'Seed-specific overexpression of phytoene synthase: increase in carotenoids and other metabolic effects', *Plant J.* 22, 401–412 (1999).

[187]J. M. Gertz, W. K. Vencil & N. S. Hill, Tolerance of transgenic soybean (*Glycine max*) to heat stress, in: *Proceedings of the 1999 Brighton Crop Protection Conference: Weeds*, vol. 3, Farham, UK, British Crop Protection Council (1999), pp. 835–840.

[188]X. Ye, S. Al Babili, A. Kloeti, J. Zhang, P. Lucca, P. Beyer & I. Potrykus, 'Engineering the provitamin A (betacarotene) biosynthetic pathway into (carotenoid-free) rice endosperm', *Science* 287, 303–305 (2000).

[189]E. Delhaize, D. M. Hebb, K. D. Richards, J. M. Lin, P. R. Ryan & R. C. Gardner, 'Cloning and expression of a wheat (*Triticum aestivum*) phosphatidylserine synthase cDNA. Overexpression in plants alters the composition of phospholipids', *J. Biol. Chem.* 274, 7082–7088 (1999).

[190]F. Murray, D. Llewellyn, H. McFadden, D. Last, E. S. Dennis & W. J. Peacock, 'Expression of the *Talaromyces flavus* glucose oxidase gene in cotton and tobacco reduces fungal infection, but it is also phytotoxic', *Mol. Breed.* 5, 219–232 (1999).

[191]S. C. H. J. Turk & S. C. M. Smeekens, Genetic modification of plant carbohydrate metabolism, in: *Applied Plant Biotechnology*, V. L. Chopra, V. S. Malik & S. R. Bhat (Eds), Science Publishers, Enfield UK (1999), pp. 71–100.

[192]K. H. Engel, G. Gerstner & A. Ross, Investigation of glycoalkaloids in potatoes as example for the principle of substantial equivalence, in: *Novel Food Regulation in the EU – Integrity of the Process of Safety Evaluation*. Federal Institute of Consumer Health Protection and Veterinary Medicine, Berlin (1998), pp. 197–209.

[193]W. Hashimoto, K. Momma, H.-J. Yoon, S. Ozawa, Y. Ohkawa, T. Ishige, M. Kito, S. Utsumi & K. Murata, 'Safety assessment of transgenic potatoes with soybean glycinin by feeding studies in rats', *Biosci. Biotechnol. Biochem.* 63, 1942–1946 (1999).

hypothesis.[194] This is not really an unexpected result because glyphosate has several well-known endocrine effects. Glyphosate and the formulated product Roundup Bioforce damage cells of human embryos and placenta in concentrations well below those recommended for agricultural use. It is therefore likely that Roundup may interfere with human reproduction and embryo development. This is not a hypothetical risk since a major metabolite of gluphosinate, 3-methylphosphinicopropionic acid (3-MPPA), as well as CryAb1 toxin, have been detected in pregnant women, their foetuses and non-pregnant women exposed to GM foods,[195] raising concern about the health effects of these molecules. Moreover, the toxic and hormonal effects of the formulations are likely to be underestimated.[196] Glyphosate-based herbicides are endocrine disruptors. In human cells, glyphosate-based herbicides prevent the action of androgens, the masculinizing hormones, at very low levels – up to 800 times lower than glyphosate residue levels allowed in some GM crops used for animal feed in the USA. DNA damage has been found in human cells treated with glyphosate-based herbicides at these levels. Glyphosate-based herbicides also disrupt the action and formation of oestrogens, the feminizing hormones.[197] Although further data is warranted, it is likely that GMOs behave as endocrine disruptors, like other widely used substances.[198]

These biochemical and metabolic changes are not always negative, nor is it yet clear what impact they may have on human health or in agronomic terms, since much uncertainty surrounds the definition of tolerance thresholds and ranges of variation of "normal" concentrations. Unintended pleiotropic effects due to transgenesis are not exclusive to biotech but were also seen to a lesser extent in varieties modified by traditional methods. This does not detract from their potential or exempt producers and regulators from assessing all the implications. It only means they cannot be assumed to be innocuous. The examples I have given are not exhaustive, but are sufficient to illustrate that we are faced with a problem that can be likened to an enormous iceberg, of which we have only just perceived the tip.

---

[194]K. Paris & A. Aris, 'Hypothetical link between endometriosis and xenobiotics-associated genetically modified food', *Gynécol. Obstét. Fertil.* 38, 747–753 (2010).

[195]A. Aris & S. Leblanc, 'Maternal and fetal exposure to pesticides associated to genetically modified foods in Eastern Townships of Quebec, Canada', *Reprod. Toxicol.* 31(4), 528–533 (2011).

[196]N. Benachour, H. Sipahutar, S. Moslemi, C. Gasnier, C. Travert & G.-E. Séralini, 'Time- and dose-dependent effects of roundup on human embryonic and placental cells', *Arch. Environ. Contam. Toxicol.* 53, 126–133 (2007).

[197]C. Gasnier, C. Dumont, N. Benachour, E. Clair, M. C. Chagnon & G.-E. Séralini, 'Glyphosate-based herbicides are toxic and endocrine disruptors in human cell lines', *Toxicology* 262, 184–191 (2009).

[198]A. M. Soto & C. Sonnenschein, 'Environmental causes of cancer: endocrine disruptors as carcinogens', *Nat. Rev. Endocrinol.* 6(7), 363–370 (2010).

6

Science fiction

We did not inherit the Earth from our fathers but borrowed it from our sons.

Cuban proverb

Ai postumi l'ardua sentenza
(Judgment is a business of our sons)

Toto'

Man spends the first half of his life ruining his health and the second half endeavouring to heal.

Leonardo da Vinci

Old and New Viruses

Crops suffer considerable damage from viruses and other microbes. It is therefore not by chance that research into resistance through genetic modifications is a priority of the biotech industry. The basic approach is to insert a sequence of viral DNA specific for a protein or protein fragment that is normally part of the viral coat (capsid) into the genome of the plant. The fact that seeds thus modified begin to synthesize part or all of that protein seems to give the plant resistance to the virus, though the reasons and exact mechanism are only partly explained.[1] Viruses are also used as vectors of genes to be inserted in host genomes for their capacity to penetrate cells of eukaryotic organisms. In one way or another, the technology for the fabrication of GMOs has continuous and direct contact with the smallest, most feared and most mutable of microorganisms.

This horizon simultaneously fascinates and terrorizes biologists since it promises results while providing glimpses of great peril. Promoting the mixing of viral genes and DNA in a host cell carries the risk of unwanted effects, such as tumours and new viruses with unknown and uncontrollable properties, potentially dangerous for crops and human health.[2]

---

[1] *Les Plantes Transgéniques en Agriculture. Dix ans d'experience de la Commission du Génie Biomoléculaire*, A. Khan (Ed.), John Libbey Eurotext, Paris (1996).
[2] S. Kuwata *et al.*, 'Reciprocal phenotype alterations between two satellite RNAs of cucumber mosaic virus', *J. Gen. Virol.* 72, 2385 (1991).

A first concern arose from insertion of the promoter gene 35S obtained from the cauliflower mosaic virus (CMV), a pararetrovirus that controls expression of many target genes and is widely used to obtain transgenic seeds. Studies on rice show that gene fragmentation and recombination often occur when using 35S but not other promoter genes.[3] This promoter is therefore an exception. Contrary to what has always been sustained, 35S can control the function of various genes in cells of animals, such as *Xenopus laevis*,[4] and in nuclear extracts of human tumour cells, such as HeLa.[5] Its stable integration into the human genome could lead to synthesis of chimeric proteins with unpredictable properties. 35S helps determine the genetic instability of transgenic cell lines[6] and also has the capacity to reactivate dormant viruses or create new ones in species in which it is inoculated,[7] causing major changes in gene expression. Despite the controversies[8,9] following the first alarm,[10,11,12] the complexity of this retrovirus is attracting much attention and prompting new research,[13] one reason being its negative effects, since viral infection of a plant species GM to resist pesticides can silence the promoter and make the plant again susceptible to herbicides. The stability of the transgene is a very important character for GM plants. Instability of the transgene can be due to various factors (e.g. too many copies inserted, rearrangement of the transgene). Functional activation may also be due to a host defence response that reacts to CMV infection by inactivating replication of the virus and blocking the transgenic sequences that have homologies with the infecting virus.[14] Since CMV is practically ubiquitous in *Brassica* crops, such as canola, it is easy to see a problem on the horizon.

---

[3]S. P. Kumpatla & T. C. Hall, 'Organizational complexity of a rice transgenic locus susceptible to methylation-based silencing', *IUBMB Life* 48, 459–467 (1999).

[4]N. Ballas, S. Brodo, H. Soreq & A. Loyter, 'Efficient functioning of plant promoters and poly(A) sites in *Xenopus* oocytes', *Nucleic Acid Res.* 17, 7891–7903 (1989).

[5]C. Burke, X. B. Yu, I. Marchitelli, E. A. Davis & S. Ackerman, 'Transcription factor IIA of wheat and human function similarly with plant and animal viral promoter. *Nucleic Acid Res.* 18, 3611–3620 (1990).

[6]M. W. Ho & R. Steinbrecher, 'Fatal flaws in food safety assessment', *Environ. Nutr. Interact.* 2, 51–84 (1998).

[7]W. M. Wintermantel & J. E. Schoelz, 'Isolation of recombinant viruses between cauliflower mosaic virus and a viral gene in transgenic plants under conditions of moderate selection pressure', *Virology* 223, 156–164 (1996).

[8]J. Hodgson, 'Scientists avert new GMO crisis', *Nat. Biotechnol.* 18, 13 (2000).

[9]M. A. Matzke, F. M. Mette, W. Aufsatz, J. Jakowitsch & A. J. M. Matzke, 'Integrated pararetroviral sequences', *Nat. Biotechnol.* 18, 579 (2000).

[10]M. W. Ho, A. Ryan & J. Cummins, 'Cauliflower mosaic viral promoter – a recipe for disaster?', *Micr. Ecol. Health Dis.* 11, 194–197 (1999).

[11]M. W. Ho, A. Ryan & J. Cummins, 'CaMV 35S promoter fragmentation hotspot confirmed and it is active in animals', *Micr. Ecol. Health Dis.* 12, 189 (2000).

[12]J. Cummins, M. W. Ho & A. Ryan, 'Hazardous CaMV promoter?', *Nat. Biotechnol.* 18, 363 (2000).

[13]M. A. Matzke, M. F. Mette & W. Aufsatz, 'More on CaMV', *Nat. Biotechnol.* 18, 579 (2000).

[14]N. S. Al-Kaff, M. M. Kreike, S. N. Covey, R. Pitcher, A. M. Page & P. J. Dale, 'Plants rendered herbicide-susceptible by cauliflower mosaic virus-elicited suppression of a 35S promoter-regulated transgene', *Nat. Biotechnol.* 18, 995–999 (2000).

The whole benefit (hoped for or hypothetical!) of transformation, mediated by promoter 35S, could be reversed by the first infection that comes along!

The specific expression of other currently used promoters (also in sectors other than agriculture) is unclear and until more is known it is premature to conclude that technologies based on DNA recombination are safe.[15] Concern about the use of viral vectors in operations of transgenesis is nevertheless reaching the threshold of alarm, due to serious "inconveniences" associated with this technique. Indeed, viral vectors are generally present in the cytosol as "extrachromosomal genomes", but in a small percentage of cases they integrate stably into the host chromosomes.[16] Despite the low frequency of integration, these intragenic insertions, chromosome deletions[17] and mutagenic or carcinogenic effects[18,19] are so important that they led to complete revision of the method of gene transfer based on viral vectors, at least for therapeutic applications.[20]

Another possibility is also emerging, namely that new viruses with unknown properties could form or be reactivated[21] as an unintended or unwanted consequence of manipulation and molecular recombination. Such a possibility is not generally considered in risk assessment procedures.[22] That aspect raises disturbing

---

[15]D. Jacob, A. Lewin, B. Meister & B. Appel, 'Plant-specific promoter sequences carry elements that are recognised by the eubacterial transcription machinery', *Transgenic Res.* 11, 291–303 (2002).

[16]H. Nakai, S. R. Yant, T. A. Storm, S. Fuess, L. Meuse & M. A. Kay, 'Extrachromosomal recombinant adeno-associated virus vector genomes are primarily responsible for stable liver transduction *in vivo*', *J. Virol.* 75(15), 6969–6976 (2001).

[17]H. Nakai E. Montini, S. Fuess, T. A. Storm, M. Grompe & M. A. Kay, 'AAV serotype 2 vectors preferentially integrate into active genes in mice', *Nature Gen.* 34(3), 297–302 (2003).

[18]Z. Li, J. Düllmann, B. Schiedlmeier, M. Schmidt, C. von Kalle, J. Meyer, M. Forster, C. Stocking, A. Wahlers, O. Frank, W. Ostertag, K. Kühlcke, H. G. Eckert, B. Fehse & C. Baum, 'Murine leukemia induced by retroviral gene marking', *Science* 296(5567), 497 (2002).

[19]E. Marshall, 'Clinical research. Gene therapy a suspect on leukemia-like disease', *Science* 298, 34–35 (2002).

[20]S. Hacein-Bey-Abina, C. Von Kalle, C. Schmidt, M. P. McCormack, N. Wulffraat, P. Leboulch, A. Lim, C. S. Osborne, R. Pawliuk, E. Morillon, R. Sorensen, A. Forster, P. Fraser, J. I. Cohen, G. de Saint Basile, I. Alexander, U. Wintergerst, T. Frebourg, A. Aurias, D. Stoppa-Lyonnet, S. Romana, I. Radford-Weiss, F. Gross, F. Valensi, E. Delabesse, E. Macintyre, F. Sigaux, J. Soulier, L. E. Leiva, M. Wissler, C. Prinz, T. H. Rabbitts, F. Le Deist, A. Fischer & M. Cavazzana-Calvo, 'LMO2-Associated clonal T cell proliferation in two patients after gene therapy for SCID-X1', *Science* 302, 415–419 (2003).

[21]A. E. Green & R. F. Allison, 'Recombination between viral RNA and transgenic plant transcripts', *Science* 263, 1423–1425 (1994).

[22]H. Hansen, M. I. Okeke, Ø. Nilssen, T. Traavik, 'Recombinant viruses obtained from co-infection *in vitro* with a live vaccinia-vectored influenza vaccine and a naturally occurring cowpox virus display different plaque phenotypes and loss of the transgene', *Vaccine* 23, 499–506 (2004).

questions and suggests horrifying scenarios: what would happen if a family with the pathogenicity of bacteria and the penetration of viruses were inadvertently created? A new season of uncontrollable pandemics may be just around the corner, considering the ease with which new viruses or even new forms of microbe life can be "generated".

According to Nobel laureate Manfred Eigen of the Max-Planck Institute for Biophysical Chemistry in Gottingen, Pandora's box is still open and new diseases are escaping from it.[23] He was referring to the emergence of new viral strains and the increased virulence of formerly innocuous bacteria. Human intervention is the first to be accused for all this. Epidemiological observers all over the world have been alerted by the powerful re-emergence of infectious diseases considered defeated, and this alarm has been fanned since discovery of the mechanisms by which new pathogenic microbes can be brought into existence. There are essentially two main ways:

1. Transcapsidation, by which a virus becomes coated in foreign proteic material, a camouflage that enables it to deceive host defences and penetrate into organisms (plants or humans) that would otherwise have rejected it.

2. Recombination, hence the technique of recombinant DNA, by which host and introduced viral DNA are remodelled to create a new viral particle. When a cell is induced to permanently tolerate a chronic infectious state, such as that due to continuous synthesis of viral proteins, the probabilities of recombination and genomic instability increase spontaneously.[24]

New viruses or viruses with unusual biological characteristics could emerge from plants rendered resistant to the same viruses, by combining heterocapsidation and recombination.[25] Gene transfer and recombination between related viral varieties can promote the development of new viral strains even under normal laboratory conditions.[26] In one case it was seen that this happened through the infamous promoter 35S obtained from CMV. Under strong selective pressure, the promoter proved able to recombine with a phytopathogenic virus in all 24 plants on which the microbe was implanted.[7] Plants engineered to resist the virus produce coat proteins (CPs) that encapsulate the viral particles, altering their properties and

---

[23] M. Eigen, La quasispecie virale, in: *Le Epidemie*, Quaderni Le Scienze (2000), 114, p. 43.
[24] R. Allison, RNA plant virus recombination, Proceedings of the USDA-APHIS/AIBS Workshop on Transgenic Virus-resistant Plants and New Plant Viruses, 1995, Beltsville, Maryland (1995).
[25] M. Tepfer, 'Viral genes and transgenic plants: what are the potential environmental risks?', *BioTechnology* 11, 1125–1132 (1993).
[26] A. E. Greene & R. F. Allison, 'Recombination between viral RNA and transgenic plant transcripts', *Science* 263, 1423–1425 (1994).

transmissibility in an unpredictable way.[27,28,29] "We do not have the data necessary to understand how such events occur under natural conditions. The modified and encapsulated virus cannot produce new CPs, so the new strains created by heterocapsidation are not propagated." Fortunately, studies are underway to limit this type of risk.[30]

However, like other characteristics coded by modified genes, these too can spread horizontally and probably more widely and faster, since they are at least in part carried by "new" recombinant viruses. The dangers associated with the emergence of new viral populations having largely unknown characteristics seem so evident that it is superfluous to underline them. Unfortunately, it is necessary to underline them because the various appeals for special caution in this domain are regularly met with silence. Let us recall some facts which will also show that *The Cassandra Crossing* is not just the title of a film.

In recent decades, microbiologists throughout the world have warned that new viruses are emerging and other formerly innocuous or only moderately pathogenic microorganisms are becoming more virulent.[31] This is primarily due to profound man-made changes in the ecology of interactions between host and pathogen and between humans and nature. Apart from the well-known case of AIDS, we can name a whole group of diseases due to microbes of the families Flaviviridae (Dengue fever in Australia, Omsk haemorrhagic fever), Bunyaviridae (Crimean haemorrhagic fever, Rift Valley fever), Arenaviridae (Lassa, Argentine, Bolivian, Venezuelan and Brazilian haemorrhagic fever) and above all Filoviridae (Marburg virus haemorrhagic fever, Ebola virus). The main causes for these viruses becoming aggressive lies in the ecological alterations caused by human activity: deforestation, contamination with animals, climate change. Environmental changes are not the only causes:

> Also development of the biotech industry has risks: preparation
> of vaccines from infected and genetically modified animal cells
> can transmit unidentified viruses to persons vaccinated. Thus a

---

[27] J. Hammond & M. M. Dienelt, 'Encapsidation of potyviral RNA in various forms of transgene coat protein is not correlated with resistance in transgenic plants,' *Mol. Plant Microbe Interact.* 10(8), 1023–1027 (1997).

[28] H. Lecoq, D. Bourdin, B. Raccah, E. Hiebert & D. E. Purcifull, 'Characterization of a Zucchini yellow mosaic virus isolate with a deficient helper component', *Phytopathology* 81, 1087–1091 (1991).

[29] D. Bourdin & H. Lecoq, 'Evidence that heteroencapsidation between two potyviruses is involved in aphid transmission of a non-aphid-transmissible isolate from mixed infections', *Phytopathology* 81, 1459–1464 (1991).

[30] C. Jacquet, B. Delecolle, B. Raccah, H. Lecoq, J. Dunez & M. Ravelonandro, 'Use of modified plum pox virus coat protein genes developed to limit heteroencapsidation-associated risks in transgenic plants', *J. Gen. Virol.* 79, 1509–1517 (1998).

[31] P. Daszak, A. A. Cunningham & A. D. Hyatt, 'Emerging infectious diseases of wildlife: threats to biodiversity and human health', *Science* 287, 443–449 (2000).

contaminated culture of monkey kidney cells led to the discovery
of a new haemorrhagic fever and a new family of viruses, the
Filoviridae.[32]

The virus in question is the agent of the terrible Ebola haemorrhagic fever, which like the Marburg virus, is of unknown origin and was only identified in 1989. A possible source of these new particles is animal cells GM and/or manipulated for use in xenotransplant programmes. The animals thus "prepared" can be excellent vectors for unknown viruses or viruses the presence of which is not actively investigated (since the usual controls are limited to analysing six groups of retrovirus and known components of the herpes family).[33] Because they are then "transplanted" under powerful immunosuppression, this may have much graver consequences than those of the rainforest microbe.[34] More generally, as underlined by the Institute of Medicine of the US government, the organs of animals used for transplants or GM to prepare transgenic derivatives (proteins, hormones, etc.) are objectively a hold-all for retroviruses, that often give rise to a recombinant virus with altered pathogenic characteristics.[35] Unexpected new virulence of a viral strain, initially "modified" to make a vaccine, was the cause of a recent scandal. Instead, the virus caused the "unexpected" death of all the animals involved in the experiment.[36] It was replied[37] that this type of transformation was predictable, but competent scientists did not assess the effects until after they were manifest.[38]

The military industry did not remain indifferent and realized the new opportunities that were offered:

> It is now possible to synthesize BW agents tailored to military
> specifications [....] it is [becoming] a simple matter to produce
> new agents but a problem to develop antidotes. New agents can
> be produced in hours; antidotes may take years. To gauge the
> magnitude of the antidote problem, consider the many years
> and millions of dollars that have been invested, as yet without
> success, in developing a means of countering a single biological

---

[32] B. Le Guenno, Ebola e altri virus, in: *Le Epidemie*, Le Scienze Quaderni, Milan (2000), 114, p. 54.

[33] R. Nowak, 'Xenotransplants set to resume', *Science* 11, 1148 (1994).

[34] J. Allan, 'Silk purse or sow's ear', *Nature Med.* 3, 275 (1997).

[35] J. Kaiser, 'IOM backs cautious experimentation', *Science* 7, 305 (1996).

[36] R. L. Jackson, A. J. Ramsay, C. D. Christensen, S. Beaton, D. F. Hall & I. A. Ramshaw, 'Expression of mouse interleukin-4 by a recombinant ectromelia virus suppresses cytolytic lymphocyte responses and overcomes genetic resistance to mousepox', *J. Virol.* 75, 1205–1210 (2001).

[37] A. Mullbacher & M. Lobigs, 'Creation of killer poxvirus could have been predicted', *J. Virol.* 75, 8353–8355 (2001).

[38] S. A. Weaver & M. C. Morris, 'Risks associated with genetic modification: an annotated bibliography of peer reviewed natural science publications', *J. Agric. Environ. Ethics* 18, 157–189 (2005).

agent outside the BW field – the AIDS virus. Such an investment surpasses the resources available for BW defence work.[39]

The speaker is Douglas Feith, US Secretary for Defence in the 1980s. We can believe his words. These new arms can be created in many ways, but the safest and most effective is certainly:

> to increase their antibiotic resistance, virulence and environmental stability. It is possible to insert lethal genes into harmless microorganisms, resulting in biological agents that the body recognizes as friendly and does not resist. Genetic engineering can also be used to destroy specific strains or species of agricultural plants or domestic animals, if the intent is to cripple the economy of a country. [...] Unlike nuclear technologies, genetically engineered organisms can be cheaply developed and produced, require far less scientific expertise, and can be effectively employed in many diverse settings.[40]

As underlined by biophysicist R.L. Sinsheimer of California University, it is impossible to distinguish peaceful and military uses of the new transgenic toxins. The International Peace Research Institute of Stockholm is also convinced of this and recalls that:

> some common forms of vaccine production are very close technically to production of BW agents and so offer easy opportunities for conversion.[41]

All this raised the concern that voluntary or accidental release of dangerous GM viruses, bacteria or fungi could cause uncontrollable pandemics. If pointing this out is being a Cassandra, then I am happy to be at the top of the list. But don't say we were not warned ....

Threats

War scenarios announced by scientific and military analysts of the 1980s were revolutionized in the last two decades, after two events: the collapse of communist countries guided by the USSR and the tumultuous progress of molecular biology that led to the new discipline of genetic engineering.

The fall of the Berlin wall and the end of the Cold War marked the end of a planetary equilibrium that paradoxically rested on that "wall". Far from being the end of the story rashly forecast by the historian Francis Fukuyama,[42] the end of the

---

[39] D. Feith, US Defence Secretary, 8th August 1986, cited in: J. Rifkin, *The Biotech Century*, Tarcher-Putnam Edition, New York (1999), pp. 92–93.
[40] J. Rifkin, *The Biotech Century*, Tarcher-Putnam Edition, New York (1999), p. 93.
[41] J. Rifkin, *Declaration of a Heretic*, Routledge & Keagan, Boston, MA (1985), p. 58.
[42] F. Fukuyama, *La Fine della Storia e l'ultimo Uomo*, Rizzoli, Milan (1992).

USSR brought out stark contradictions and tensions that lose their regional character in the era of economic globalization, threatening to become global. In a context of increasing disparity between rich and depressed countries, even "local" conflicts end up involving economic, ideological and religious levels, not expressing as conventional war but promoting acts of systematic terrorism that are now an alternative military strategy. There are many convincing reasons why it is not surprising that terrorist groups and organizations, autonomously or somehow manoeuvred by "mongrel states", find bacteriological weapons particularly convenient.[43]

Like the opening of Pandora's box, the fall of the USSR and other dictatorial regimes (like apartheid) and the consequent *diaspora* of technologies and scientists once occupied in the sector of non-conventional warfare has created a virtually uncontrollable market where it is relatively easy to acquire the know-how and basic resources for building any bacteriological weapon. In parallel, progress of genetic engineering has made it possible to construct biological vectors capable of inflicting irreversible and lethal damage.[44] Bacteria, fungi and viruses can be "engineered" to increase virulence and infectiousness, making them resistant to antibiotics or insensitive to vaccines. The costs of such operations and the associated risks are infinitely lower than those of conventional and thermonuclear weapons. The procedures for release are much simpler and there are many opportunities for spread. This is not just a possible scenario, but something much more concrete, already widely implemented in the ex-USSR, as documented by a large group of scientists (Pasechnik, Popov, Alibek), previously involved in the bacteriological warfare programme known as "Biopreparat",[45] who escaped to the West.

The Soviet programme, that employed more than 25,000 technicians and scholars in 28 centres, had already produced the first GM microbe in 1983: a hyper-virulent strain of *Francisella tularensis*, the aetiological agent of tularaemia. Over the years, modified strains of smallpox (in which the gene for Venezuelan equine encephalitis, beta-endorphin, and the notorious Ebola virus had been inserted) and especially varieties of anthrax (*Bacillus anthracis*), resistant to vaccines and antibiotics, were produced. Other more sophisticated programmes (such as the Bonfire and Factor projects) were aimed at engineering *Yersinia pestis* (the agent of bubonic plague) to express diphtheria toxin or a pathogenic viral factor, paradoxically produced by the bacillus only after it was exposed to antibiotic treatment. The list of these criminal "discoveries" is much longer and perhaps still not completely

---

[43]M. R. Hilleman, 'Overview: cause and prevention in biowarfare and bioterrorism', *Vaccine* 20, 3055–3067 (2002).

[44]J. R. Gilsdorf & R. A. Zilinskas, 'New considerations in infectious disease outbreaks: the threat of genetically modified microbes', *Clin. Infect. Dis.* 40, 1160–1165 (2005).

[45]L. Szinicz, 'History of chemical and biological warfare agents', *Toxicology* 214, 167–181 (2005).

known but reveals the entity of the threat we were under.[46] Production of new high-virulence microbes is a danger against which no country can yet defend itself in a satisfactory manner. Were we to be faced with an epidemic caused by a new vector, it would be difficult to recognize the cause, differentiate the clinical characteristics and implement effective mass measures in time. By definition, diseases caused by GM microbes have unusual clinical pictures and courses, and the aetiological agent cannot be determined by routine diagnostic tests. Vaccines in current use do not work, since the modified strains express different antigens or may be modified to produce proteins that block host immune response. Finally, the new vector is often resistant to commonly used antibiotics or reacts to them by producing a new toxin. Obviously, the USSR was not the only country to conduct research in this direction, but it is the only country, besides Iraq (mainly anthrax) and South Africa (until the end of apartheid), on which we have convincing and articulated documentation. Other countries actively involved in research in this sector are the USA, Iran, China, North Korea and probably Israel and Cuba.[47]

The use of techniques of molecular transgenesis has not only revolutionized the logic and "classical" approach of bacteriological warfare but has also suggested unimaginable applications[48] and made new measures of prevention and control necessary.[49]

The completely new and unexpected scenario of "food bioterrorism" recently overlapped with the classical concept of bacteriological attack. Modified plant or animal genes can cause selective, lethal zoonoses or destroy strategic crops (wheat, rice, soy), spreading engineered seeds to express antinutritional characters or equipped with "terminator" genes that make the progeny sterile. It is also possible to spread herbicide-resistant weeds or GM parasites. In one way or another, the characteristics expressed by genetically transfected seeds can damage agricultural production of crops and animals. The result is not only economic damage but also political and social destabilization, which could be severe for developing nations and those still largely dependent on a few vulnerable animal or crop species (for example rice in the diet of Asian countries). This offensive strategy has many advantages: (a) the agent need not be dangerous for humans or for those who implement the attack; (b) preparation of the vector requires little technology and is inexpensive; (c) systems for spreading and releasing the agent are unsophisticated and accessible; (d) dispersal in a few distant points can be sufficient for effective

---

[46]E. J. DaSilva, 'Biological warfare, bioterrorism, biodefence and the biological and toxin weapons convention', *Electr. J. Biotechnol.* 2(3), 99–129 (1999).

[47]J. Guillemin, 'Scientists and the history of biological weapons', *EMBO J.* 7, S45–S49 (2006).

[48]C. Dennis, 'The bugs of war', *Nature* 411, 232–235 (2001).

[49]C. M. Fraser & M. R. Dando, 'Genomics and future biological weapons: the need for preventive action by the biomedical community', *Nat. Genet.* 29, 253–256 (2001).

spread; (e) the attack can go unnoticed for a long time and be ascribed to natural causes. Eradication of an agricultural epidemic calls for immediate action to confine and arrest spread of the pathogen at the outset.[50] However, as documented by many episodes of natural infestation, years usually pass before the threat is recognized as such and is correctly diagnosed.[51] These features make agricultural bioterrorism a real threat for certain states, perhaps much more than an attack on the civilian population.[52]

Programmes aimed at striking the agriculture of the adversary have long been part of the arsenal of states armed with chemical and bacteriological weapons,[53] but with the growth of the biotech industry and advances in molecular biology, GM organisms have recently emerged as an alternative form of total warfare, much more dangerous than any other form, including nuclear.[54,55] A completely new strategy for using these vectors is to identify an agricultural sector as target. Programmes based on bacteriological weapons against plants and food have been known since the end of the Second World War, though they have only taken on a global dimension with advances in genetic engineering.[56] The spread of zoonoses and/or infestation of cereal crops may have both economic and socio-political repercussions on the stability of a nation. Food shortages lead to increased prices and unemployment. Countries with agriculture based prevalently on extensive monocultures and selective raising of a single animal species are particularly vulnerable. The absence of a rigorous system of monitoring the whole food production line, an aim that can be achieved only using an efficient model of traceability covering every stage from producer to consumer, makes it impossible to adopt fast and effective countermeasures. These features mean that ideal target nations are rich countries like USA and China, where monoculture and the lack of any efficient regulation of production are the rule. An attack against the food production of these nations would trigger a cascade of events, amplifying its destructive impact. Damage to crops and animals caused by an epidemic would almost automatically close international markets, creating new opportunities for exporters hitherto excluded. The recent saga associated

[50]M. R. Finckh & M. S. Wolfe, Diversification strategies, in: *The Epidemiology of Plant Diseases*, D. G. Jones (Ed.), Kluwer, Dordrecht, the Netherlands (1998), pp. 231–259.

[51]T. S. Schubert, S. A. Rizvi, X. Sun, T. R. Gottwald, J. H. Graham & W. N. Dixon, 'Meeting the challenge of eradicating citrus canker in Florida', *Plant Dis.* 85, 340–356 (2001).

[52]M. Wheelis, *Agricultural Biowarfare and Bioterrorism*, Pugwash Study Group on the Implementation of the Chemical and Biological Weapons Convention, Switzerland (2000), ISBN 1-930169-14-0.

[53]M. Wheelis, Biological sabotage in World War I, in: *Biological and Toxin Weapons: Research, Development and Use from the Middle Ages to 1945*, E. Geissler & J. E. v. Moon (Eds), Oxford University Press, Oxford (1999), pp. 35–62.

[54]D. A. Henderson, 'The looming threat of bioterrorism', *Science* 283, 1279–1282 (1999).

[55]R. A. Zilinskas & B. K. Zimmermann, *The Gene Splicing Wars. Reflections on the Recombinant DNA Controversy*. Macmillan, New York (1986).

[56]S. M. Whitby, *Biological Warfare Against Crops*, Palgrave, Hampshire, UK (2002).

with BSE, in which Great Britain and other affected countries (including the USA) could no longer export their beef, demonstrates the losses and serious effects of this scenario. The case of BSE was fateful, since the prion was not deliberately spread, but the future could hold many surprises. The use of GM vectors (plants, insects and microbes) to damage the economies of other states and manipulate markets and our future is already being contemplated.

The character of available ways of attacking livestock and agriculture are not known. The potential lethality of such attacks is illustrated by the costs of mad cow disease (BSE). In the USA, agricultural terrorism could have a significant impact on the economic well-being of the country because agriculture accounts for 13% of the US domestic product. Agricultural terrorism may not have the shock value of direct attacks on humans, but to terrorists it offers many other advantages: it is virtually undetectable, it could be unattributable and it might never be understood as a deliberate attack although the effects could be lasting and nationwide.[57]

This is not science fiction, but a real scenario with concrete examples. A few years ago, two programmes for the production of plant pathogens classified as "bio-control agents" were extensively financed by the USA and Great Britain.[58] The first aimed to produce fungal spores of *Fusarium oxysporum*[59] specialized in attacking the coca plant, and the second was the fungus *Pleospora papaveracea*,[60] developed in laboratories in Uzbekistan, to destroy poppies. The two projects were sponsored by the United Nations Drug Control Program with the aim of striking a mortal blow to drug traffic. Unfortunately, open opposition by the governments of the countries where the microbes were to be released, together with mobilisation of political and green activists,[61] prevented the operation and induced the United Nations to withdraw its consent. The episode documents the distance between official declarations and the real intentions of certain countries to fight drug abuse. Above all, it demonstrates how operations of molecular surgery can be organized today in order to selectively strike agricultural production in well-defined geographical areas. According to M. Wheelis, an analyst of the Pugwash Study Group:

> Although there is no evidence that the agents are being developed for hostile use, the absence of target country approval

---

[57]A. H. Cordesman, *The Challenge of Biological Terrorism*, CSIS Press, Washington, DC (2005).

[58]M. Jelsma, *Vicious Circle; The Chemical and Biological War on Drugs*, Transnational Institute, Amsterdam, available at: www.tni.org/drugs/.

[59]W. J. Connick Jr, D. J. Daigle, A. B. Peppermann, K. P. Hebbar & R. D. Lumsden, 'Preparation of stable, granular formulations containing *Fusarium oxysporum* pathogenic to narcotic plants', *Biol. Control* 13, 79–84 (1998).

[60]A. Bailey, P. C. Apel-Birkhold, N. R. O'Neill, J. Plaskowitz & S. Alavi, 'Evaluation of infection processes and resulting disease caused by *Dendryphion penicillatum* and *Pleospora papaveracea* on *Papaver somniferum*', Phytopathology 90, 699–709 (2000).

[61]K. Kleiner, 'Operation eradicate', *New Sci.* 163(2003), 20 (1999).

> makes it equally difficult to demonstrate that they are being developed for peaceful purposes. This ambiguity raises legitimate concerns about compliance with Article I of the BTWC. Furthermore, once effective agents have been developed, the intense concern over the drug trade in drug-consuming states may lead to pressure to use them covertly, regardless of target country approval.[54]

Despite the good intentions of the proponents, we shall never know whether that type of measure would have annihilated drug production. However, its application would probably have caused a disaster: destruction of poppy fields would also have destroyed the farm economies and local communities that depend on the poppy harvest and would have profoundly altered the ecosystems involved, with unpredictable consequences. Vectors of cereal and other plant diseases can infect and proliferate in other species. The genetic diversity of local plant varieties used for food and traditional animal raising could be gravely damaged by intentional or accidental release of such biological weapons.[62] There is also a paradox: how could a programme of biological control of coca and other pharmacological plants be effective when the normal programmes of biological monitoring of weeds have encountered so many difficulties in the actuation?[63] All this brings to mind a medieval adage: "The road to hell is paved with good intentions."

Recent history offers other examples on the military or criminal use of selected or modified organisms to damage food production. In the eighties and nineties, Iraq possessed two strains of wheat cover smut, *Tilletia tritici* and *Tilletia laevis*, prepared for use in the war against Iran. The project was conceived in 1974, but it was only with inspections after the Gulf War of 1991 that it was discovered how close the Iraqis were to making the agent a true weapon and to fully appreciating its intrinsic peril.[64] More recently, a group of farmers illegally introduced rabbit haemorrhage disease virus (RHDV) into New Zealand.[65,66]

The damage that would be caused by an attack designed to alter food production ecosystems is difficult to quantify, but is certainly large. It includes direct costs that

---

[62]J. P. Dudley & M. H. Woodford, 'Bioweapons, bioterrorism and biodiversity: potential impacts of biological weapons attacks on agricultural and biological diversity. *Rev. Sci. Tech. Off. Int. Epiz.* 21(1), 125–137 (2002).
[63]L. V. Madden & M. Wheelis, 'The threat of plant pathogens as weapons against US crops', *Annu. Rev. Phytopathol.* 41, 155–176 (2003).
[64]S. M. Whitby & P. Rogers, 'Anticrop biological warfare – implications of the Iraq and US programs', *Def. Anal.* 13, 303–318 (1997).
[65]J. Henning, C. Heuer & P. R. Davis, 'Attitudes of New Zealand farmers to methods used to control wild rabbits', *Prev. Vet. Med.* 67(2–3), 171–194 (2005).
[66]N. L. Forrester, B. Boag, S. R. Moss, S. L. Turner, R. C. Trout, P. J. White, P. J. Hudson & E. A. Gould, 'Long-term survival of New Zealand rabbit haemorrhagic disease virus RNA in wild rabbits, revealed by RT-PCR and phylogenetic analysis', *J. Gen. Virol.* 84, 3079–3086 (2003).

can be ascribed immediately to loss of food stocks or animals, and indirect costs which are probably several orders of magnitude larger. Contamination of a crop or infection of only a few animals involves costs to eradicate and contain the epidemic, lost exports and import restrictions for quarantine. A few examples illustrate the catastrophic nature of this possibility.

In 1997, the epidemic of foot and mouth disease (FMD) in Taiwan caused direct costs of several hundred million dollars; the costs to contain the epidemic were more than 4 billion and lost exports reached 15.5 billion dollars. In 1993, FMD in Italy had only marginal direct costs, eradication costs exceeded 12 million dollars and lost exports were over 120 million. In 2001, the same epidemic in England, probably introduced unintentionally by illegal import of infected meat, cost about 50 billion dollars in lost exports.[67]

The future, however, holds even more perverse and malignant biotech applications. The techniques used to create GM seeds are sources of fantastic new solutions for many a young Dr Strangelove:

> Terminator technology that renders seed infertile to guarantee seed corporations' yearly sales may eventually be abused for economic warfare. If terminator crops become widespread, it would be easy for a transnational company that controls the technique to stop sales to a specific country or region for political or economic purposes. After some years of planting such seeds, only limited quantities of other seed would be available, thus agriculture could be paralyzed, leading to serious economic crisis and/or famine.[68]

Worse still, current state of the art makes it possible to develop microbes, insects and other predators to destroy one or more varieties of plants or animals, selected for their geographical specificity and economic/political importance: spread of these vectors would bring food production of any target country to its knees within years or months. It would be possible to manipulate old and new microbes, develop modified seeds to express one or more toxic proteins (like the pesticide properties of Bt seeds) or even insert gene sequences from pathogenic viruses into seeds, so that their functions could be reactivated once eaten. The scenario is horrifying but not impossible: many other vectors modified genetically to do harm will follow, even if they are inconceivable in the present perspective.[69] The American Society for Microbiology coded a long series of rules to identify and label agents and substances that could be used by terrorists. Among these, GMOs are in category F: GMOs are generally first- and second-class agents of infection (Category A

---

[67] V. Gewin, 'Agriculture shock', *Nature* 421, 106–108 (2003).
[68] The Sunshine Project, An introduction to biological weapons, their prohibition, and the relationship to biosafety, available at: http://www.sunshine-project.org/.
[69] The Sunshine Project, *op. cit.*, April 2002, available at: http://www.sunshine-project. org/publications/bk/pdf/bk10en.pdf.

[UN2814 or UN2900], Category B [UN3373]) or else substances or organisms in class 9 (miscellaneous dangerous goods).[70] The text specifies that GMOs are organisms, the genome of which has been modified or altered intentionally by genetic engineering in ways that could not occur naturally, and must be classified like potentially infectious substances and diagnostic or clinical samples.

What to Do?

To prevent the spread of plants and foods modified genetically in order to provoke economic damage or to harm health, it is first necessary to pass laws to enable products to be traced, to reconstruct the production line from seed to shop shelf. Similar provisions are being discussed in the European Parliament to assure consumers about the absence of GMO components in food, irrespective of how they come to be there. The indication, aimed at ensuring greater transparency of the processes involving the transgenic industry today, was requested by analysts concerned about illegal traffic of biotech material and especially its use in warfare. The interdependence of the food market means that an attack in any region quickly becomes global, and hence a global security system is needed to share research, information and strategies. Thus in 2000, experts in the sector asked that the Cartagena Protocol on Biosafety (CPB) (to assess the ecological impact of gene transfer between species) be coordinated with the Biological and Toxin Weapons Convention (BTWC) in order to share control systems and ensure efficient exchange of information.

Unfortunately, on this very aspect the USA expressed firm opposition. Since 1993 when recombinant DNA methods were developed, the indicted technologies have been fundamentally of dual-use: the same procedures and materials (reagents, enzymes, etc.) can be used for civil and military purposes and it is often difficult to distinguish the two. Thus signatories of the CPB charged a working group to develop a protocol to protect biodiversity, specifically orientated to study the spread between species of any living GMO produced by modern biotech, which could have adverse effects on the conservation and sustainable use of biological diversity. GMOs (seeds, insects, viruses) come under this definition, considered from the specific perspective of protecting human life. The aims of the CPB (articles 1, 2 and 18) are to help prevent or reduce the risks, especially those to human health. The CPB embraced the precautionary approach established by principle 15 of the Rio Declaration on Environment and Development, according to which "lack of full scientific certainty shall not be used as a reason for postponing cost-effective measures to prevent" potential adverse effects. In this framework

---

[70]American Society of Microbiology, Sentinel laboratory: guidelines for suspected agents of bioterrorism, in: *Biological Safety: Principles and Practice*, 4th Edition, D. O. Fleming & D. L. Hunt (Eds), ASM Press, Washington, DC (2006).

it was proposed that GM plants produced for cultivation and that could therefore upset ecosystem equilibrium should be transported and used only if labelled with full details of biological characteristics and the type of genetic modification made, in order (theoretically) to permit GMOs to be traced and segregated. Since the CPB and BTWC concern biological agents, modified and otherwise, in relation to their possible impact on biodiversity and human health, it was suggested that the two should be brought into line with each other: the negotiation of the two conventions follows similar paths, and this parallel extends to problems related to normative alignment, ratification of treaties, technical and scientific methods of identifying and monitoring agents and exchange of information.[71]

When the USA realized that ratifying the BTWC, including aspects related to the spread of GMOs, meant making certain commercial and technological operations transparent, they pulled out to protect industrial secrets and also certain military bacteriological programmes, where the line between defence and offensive potential was mere sophistry. However, 11th September 2001 and the increasing international threat of terrorist attacks were sources of deep reflection. Analysis conducted by scientists and specific government commissions showed that US food production was particularly vulnerable to terrorist attack.[72,73] The conclusions were unequivocal: the USA was not adequately prepared to face or prevent an attack on agriculture.[74] This bitter statement induced the Bush administration to take new measures under the presidential directive on Homeland Security HSPD-9 of 30th January 2004 to protect the agricultural economy of the nation. In particular:

> At the same time the US regulatory infrastructure for food safety is still a work in progress and is hobbled by over dependence on the private sector and underdependence on international cooperation. Whether it is a matter of detection surveillance or information flow, the US government is currently dependent on the private sector for cooperation and support [...] this dependence on the private sector is burdensome for companies and both insufficient and unreliable for ensuring the public's food safety concerns [...] FDA and the *Custom & Border Protection Agency* (Washington, DC, USA) still have not adequately funded the enforcement infrastructure nor trained personnel to ensure statistically random, uniform inspections under the new

---

[71]F. Mauro, Possible linkages between the Cartagena Biosafety Protocol and the Biological Toxic Weapons Convention. *Convention Conference Paper for Biosecurity and Bioterrorism*, Istituto Diplomatico Mario Toscano, Rome 18–19 September 2000, available at: http://lxmi.mi.infn.it/~landnet/Biosec/mauro.pdf.

[72]T. W. Frazier & D. C. Richardson (Eds), *Food and Agricultural Security: Guarding Against Natural Threats and Terrorist Attacks Affecting Health, National Food Supplies and Agricultural Economics*, New York Academy of Science, New York (1999).

[73]National Research Council (NRC), *Countering Agricultural Bioterrorism*. NRC, National Academy Press, Washington, DC (2002).

[74]J. Mervis & E. Stokstad, 'NAS Censors Report on agriculture threats', *Science* 297, 1973–1974 (2002).

pre-notification time frames [...] food protection and terrorism prevention have to be internationalised particularly given the advances that Europe and Japan have achieved in this regard.[75]

The article went on to denounce the limits and inefficiency of the system of control operating in the USA and recalled, among other things, the case of BSE infection in 2004, which confirmed that prompt intervention could not be ensured due to deficiencies of the current system of monitoring and control. Surprisingly, among the various recommendations there was investment in the creation of foods modified genetically to resist pathogens. Clearly they did not realize that introduction of these varieties would increase the fragility of the system, because, as they admit, the necessary monitoring and protection of biodiversity are not ensured and because it would open the way to the spread of GMOs manipulated for the purpose of terrorist attack. In the words of S. Wuerthele:

> It might be productive for the biotech industry to first consider how genetically modified crops themselves could contribute to terrorist attacks. Just a few bushels of "pharmacorn" producing a swine vaccine could, if strategically planted by terrorists, contaminate virtually the entire US corn supply and close international markets to us for years. [...] before genetically modifying all of our crops and animals there are simpler and more obvious steps that should be taken to protect the food supply.[76]

Indeed, it is evident that just as plants can be engineered to express a toxin or to confer specific protection, they can also be modified to produce toxins or incapsulate gene segments of viruses that would reactivate once ingested. These things are within the competence of modern labs and of terrorist groups sufficiently organized and financed. The manufacture of offensive GMOs would of course require tests and trials like other GMOs. This implies longer development times and coordination of activities that for the moment make them less attractive and less economical.[77] In other words, the probability that terrorists could decide to build a GM seed as vector for a bacteriological attack against agriculture and humans is currently only a *relative* impossibility. As situations and opportunities change, it could soon become possible, and according to many experts, we will shortly be faced with a scenario of bioterrorism.[78] Some nations, perhaps those that have been most vocal against chemical and bacteriological attack, may already be acting, planning their own protection in case they are threatened by the USA.

---

[75]R. Gilmore, 'US food safety under siege?', *Nat. Biotechnol.* 22, 1503–1505 (2004).

[76]S. Wuerthele, 'Pharmacrops and bioterror', *Nat. Biotechnol.* 23, 170 (2005).

[77]R. A. Zilinskas, Possible terrorist use of modern biotechnology techniques. *Convention Conference Paper for Biosecurity and Bioterrorism*, Istituto Diplomatico Mario Toscano, Rome 18–19 September, 2000, available at: http://lxmi.mi.infn.it/~landnet/Biosec/zilinskas.pdf.

[78]R. A. Norton, 'Food security issues: a potential comprehensive plan', *Poultry Sci.* 82, 958–963 (2003).

In Anglo-Saxon countries, this new threat has aroused lively debate about the ethics and lawfulness of scientific information about dual-use technologies published in specialist journals.[79,80,81,82] In our opinion, any attempt to limit the spread of scientific knowledge would encounter insurmountable practical difficulties, before considering ethical or legal obstacles: in a globalized world where communication is easy and instantaneous, where scientists are in continuous contact and where molecular biology labs have developed almost everywhere, the idea of containing or prohibiting exchange of scientific information is an illusion. Another road must be taken. First the BTWC must be implemented, organizing points in common with the CPB: if the international community cannot implement these two instruments, it will be unable to inspect, impose regulations and apply sanctions. Unfortunately, many countries, including the USA, have not ratified these treaties. An attack on agriculture using seeds, agents of disease or GM plants is likely to go unobserved for a long time and only be recognized when it is too late. It is necessary to become equipped with tools for epidemiological detection and molecular screening that can compare modified and/or suspicious gene sequences with those in a database which does not yet exist. The new technologies available today, such as metabolomics, in synergy with an "omics" approach (genomics and proteomics), could quickly provide the fingerprint of any suspicious species if properly applied.[83] The capacity to identify and face the threat would be a deterrent in itself. Clearly, to equip with efficient technological infrastructure of control would require laws imposing more stringent rules and the possibility of monitoring the whole food production chain: food must be properly labelled so that it can be traced through all stages of production. All this does not remove the Achilles heel of many modern agricultural economies, namely monoculture.

> Agriculture, particularly in many developed countries, has several properties that make it vulnerable to attack with genotype-specific weapons. Typically, it employs monocropping of large acreages with genetically identical cultivars, and high-density husbandry of genetically inbred animal strains. These agronomic practices reduce the genetic variability that makes populations resistant to genotype-specific weapons, and thereby create conditions (large, dense populations) that facilitate disease spread.[54]

Such decisions require a radical change in agricultural and economic policy, implying a profound change in the dominant paradigm and suggesting a direction opposite to that in which proponents of GMOs are pushing. Are politicians and the public aware of this? Or will it take a disaster to make us aware?

[79] J. Couzin, 'A call for restraint on biological data', *Science* 297, 749–750 (2002).
[80] 'Statement on the consideration of biodefence and biosecurity', Editorial, *Nature* 421, 771 (2003).
[81] 'Risks and benefits of dual-use research', Editorial, *Nature* 435, 855 (2005).
[82] 'Biosecurity with "bio-sense" ', Editorial, *Nat. Immunol.* 5, 1191 (2004).
[83] O. Fiehn, 'Metabolomics – the link between genotypes and phenotypes', *Plant Mol. Biol.* 48, 155–171 (2002).

7

Risks, rules and precautions

> The use of the precautionary principle was criticized by some as unscientific in this context. In fact, the intrinsic problem of knowability, posed by the biological complexity of the problem, makes the use of precautionary decision making particularly suitable in this arena. The assumption that plausible dangers are negligible, even when it is known that such dangers are constitutively very difficult to measure, may be more unscientific than the use of precaution.[1]
>
> *David L. Smith*

> I never think about the future, it arrives so soon.
>
> *Albert Einstein*

Forgotten Risks

We all remember the case of thalidomide.[2] The drug was released in the early 1960s and was successfully prescribed for migraine. Among those who took the drug were many pregnant women. There followed a world epidemic of terrible neonatal malformations. Only with great difficulty, after sporadic reports in *The Lancet*, was this disaster ascribed to the use of thalidomide, which incidentally had passed the necessary experimental tests. Unfortunately the tests did not consider that the drug could be prescribed for pregnant women and so tests on pregnant mice were not conducted. When these tests were finally done, it became evident that the compound was teratogenic and it was immediately withdrawn from the market. A bitter lesson seemed to have been learnt, as the legislation of industrialized countries subsequently included a long and complex testing procedure before a new drug could be sold. In actual fact, this approach has been profoundly revised, almost in silence, in the last few years. For example, AZT, an elective drug for treating AIDS, was authorized in the absence of demonstrated efficacy

[1]D. L. Smith, J. Dushoff & J. G. Morris Jr, 'Agricultural antibiotics and human health', *Plos Med.* 2(8), 1–5 (2005).
[2]S. V. Rajkumar, 'Thalidomide: tragic past and promising future', *Mayo Clin. Proc.* 79, 899–903 (2004).

though its toxicity is widely documented. The criticism expressed by Nobel laureate Kary Mullis was fruitless.[3] Vested interests prevailed over scientific rigour and common sense.

Something similar is happening for transgenic foods, which have been considered safe in the absence of any compelling evidence. This is not surprising because it reflects the unrealistically optimistic attitude of government authorities and their sanctimonious illusions about the progress that biotechnologies will "undoubtedly" bring, a certainty shared by few and not confirmed by science. The US Department of Agriculture allocates less than 1% of its risk assessment budget to the activities and products of the biotech industry, despite all the controversies of recent years. Including the funds allocated by several universities, less than three million dollars are spent per year to examine the entire spectrum of risks connected with environmental release of transgenic plants and animals.[4] However, it must be highlighted that, contrary to claims by the GM industry and its supporters, the FDA has never approved any GM food as safe. Instead, it de-regulated GM foods, ruling that they are substantially equivalent to their non-GM counterparts and do not require any special safety testing. The term "substantial equivalence" has never been scientifically or legally defined, but is used to claim (incorrectly) that GM foods are no different from their non-GM counterparts. The FDA's ruling was widely recognized as an expedient political decision with no basis in science. More controversially, the FDA ignored the warnings of its own scientists that genetic modification is different from traditional breeding and poses unique risks to human and animal health.[5]

Be as Wise as Serpents and as Simple as Doves[6]

Release of GMOs into the environment and their use as human food raises many concerns about the impact they may have on ecological equilibria and on the health of all forms of life. Assessment, evaluation and control are currently still the subject of much controversy. The first modified varieties, mainly soy, maize, cotton and rapeseed, were marketed in the absence of any rigorous laws and regulations, which were only pieced together *a posteriori* by a process of correction and addition. Available scientific data produced by companies or public research centres are still limited: for example, there have been no long-term studies to assess ecological and health sustainability in the face of increasingly frequent reports of real or potential

[3]K. Mullis, *Dancing Naked in the Mind Field,* Vintage, New York (2000).
[4]A. A. Snow & P. M. Palma, 'Commercialization of transgenic plants: potential ecological risks', *Bioscience* 2, 94 (1997).
[5]FDA documents, including statements from FDA scientists on the risks of GM foods, available at: http://www.biointegrity.org/list.html.
[6]Matthew 10: 16.

adverse effects.[7] Strong doubts have been raised about the secondary effects of GMOs on non-target organisms and about the danger of gene transfer. Secondary consequences of transgene insertion can have unexpected effects, producing toxins, amplifying the mutation rate and synthesizing allergenic proteins. Many of these dangers can be checked beforehand in the laboratory or after field trials, but not always: in some situations the risk seems to be unpredictable and can only be discovered afterwards. Faced with this spectrum of negative possibilities, the absence of reliable information dictates adoption of the Precautionary Principle (PP). The PP is based on awareness of the scientific uncertainty surrounding the negative effects of a phenomenon, product or process, and expresses this uncertainty by demanding exhaustive risk assessment. The latter involves identification, determination and management of potential danger so that the whole of society, not just science and industry, is consulted and asked to determine what threshold, if any, is acceptable for a given risk. On this point, public opinion and the scientific community are both divided.[8]

The PP appeared in European environmental policy at the end of the 1970s, but in the last 15 years it has been assimilated into national legislation and international treaties, in an increasingly authoritative way, as a basic element of ecological management.[9] Its validity was confirmed by inclusion in the Declaration of Rio:

> In order to protect the environment, the precautionary approach shall be widely applied by States according to their capabilities. Where there are threats of serious or irreversible damage, lack of full scientific certainty shall not be used as a reason for postponing cost-effective measures to prevent environmental degradation.[10]

These considerations were ratified by the EU,[11] which emphasizes in its documents the eminently "political" nature of the criteria for application of the PP, and thus by the CPB which produced a version implemented as appendix of the Convention on Biological Diversity (CBD).[12] In that context it was immediately clear that the

---

[7]L. L. Wolfenbarger & P. R. Phifer, 'The ecological risks and benefits of genetically engineered plants', *Science* 290, 2088–2093 (2000).

[8]A. I. Myhr & T. Traavik, 'The precautionary principle applied to deliberate release of genetically modified organisms (GMOs)', *Microb. Ecol. Health Dis.* 11, 65–74 (1999).

[9]D. Freestone & E. Hey, Origins and development of the precautionary principle, in: *The Precautionary Principle and International Law*, D. Freestone & E. Hey (Eds), Kluwer Law International, the Netherlands (1996).

[10]Article 15 of the Declaration of Rio. The complete text is available at: http://www.unep.org/Documents.Multilingual/Default.asp?DocumentID=78&ArticleID=1163.

[11]EU, Commission of the European Communities COM, *Communication on the Precautionary Principle*, Brussels, February 2000, COM1, available at: http://eur-lex.europa.eu/LexUriServ/site/en/com/2000/com2000_0001en01.pdf

[12]CBD Cartagena Protocol on Biosafety (2000) available at: http://bch.cbd.int/protocol/.

PP shares much of the inspiration of the concept of sustainable development,[13] an organizational principle around which strategies and environmental policies opposing globalization have been elaborated. This is another reason why application of the PP immediately ran up against a variegated series of obstacles. Some were intrinsic and mainly regarded the lack of clear applicative guidelines,[14] which led to ambiguous formulations, confused interpretations and many political decisions ranging from moratoria to simple restriction on the use and circulation of goods, especially GMOs.

Other problems were political and economic. With few exceptions,[15] the PP was openly opposed by US government authorities and regulatory bodies, headed by the FDA. Indeed, the founding criterion of the US legislation says that a product can circulate freely on markets unless it is demonstrated to be unsafe. This is diametrically opposite to the PP which says that goods can only be marketed after determination of their safety. Clearly the uncertainty implied by this approach could paralyse all activity if applied indiscriminately. Because it does not envisage correct characterization of risk, the PP, extensively interpreted, could provide a convenient alibi to opponents of any type of change. The USA considers the principle a scam to avoid World Trade Organization (WTO) agreements, especially when invoked by European authorities in support of policies affecting market freedom of other countries (not necessarily only the USA).[16]

> The Precautionary Principle is a nebulous doctrine invented by Europeans as a means to erect a trade barrier against any item produced more efficiently elsewhere.[17]

Many members of the scientific community have expressed opposition to the PP. The mixing of heterogeneous conceptual categories and scientific, political and

---

[13]M. Khor, *Proprietà intellettuale, biodiversità e sviluppo sostenibile*. Baldini Castoldi Dalai, Milan (2004).

[14]K. R. Foster, P. Vecchia & M. H. Repacholi, 'Science and the precautionary principle', *Science* 288, 979–981 (2000).

[15]One exception is reformulation of the PP in line with the Wingspread definition, proposed by a group of American scientists in 1997, according to which all necessary precautionary measures should be taken when an activity may threaten health or the environment, even if some cause–effect relationships have not been established scientifically (C. A. Raffensberger & J. Tichner (Eds), *Protecting Public Health and the Environment: Implementing the Precautionary Principle*, Island Press, Washington, DC (1999)).

[16]For example, economists of the World Bank have stated that rigorous application of the PP with adoption of new standards proposed for aflatoxins would lead to a two-thirds reduction in European imports of cereals and nuts from poor African nations (G. Majone, 'The precautionary principle and its policy implications', *J. Common Market Studies* 40, 89–110 (2002)).

[17]B. Goldstein & R. S. Carruth, 'The precautionary principle and/or risk assessment in World Trade Organization decisions: a possible role for risk perception', *Risk Anal.* 24(2), 491–499 (2004).

economic orders in interpreting the PP aroused vehement discussion,[18,19,20] particularly with regard to the nature of the relation between the PP and science-based risk assessment. According to many scientists, it was an error to suspend a process or decline to act, without certain proof of harm and a cause–effect relationship: the approach was labelled as anti-scientific, irrational, expensive and full of social, economic and cognitive consequences.

There is certainly a gap between perception of potential risk and the scientific evidence necessary to prove it. This gap may never be bridged and if the probability of risk remains aleatory, it will condemn us to inaction and block programmes on the basis of considerations that have nothing scientific about them, raising questions of a different nature: political, economic, religious and cultural. Some critics have correctly pointed out that the PP does not expressly mention the tolerance threshold beyond which a risk cannot be accepted by a given society. This threshold depends on *perception* of the problem, which is influenced by culture and tradition, and on cost–benefit assessments regarding introduction of a new technology. Unfortunately, the intrinsic reversible/irreversible nature of risk (surely an important parameter from the scientific point of view), in practice ends up being a variable subordinate to the former, at least in the current debate on the impact of biotechnologies. Other types of industrial processes and biotech applications (recombinant drugs, blood transfusions, airbags for cars, nuclear magnetic resonance) with undeniable quantifiable risks have been passed without any doubts or contestations. Those disputing the validity of the PP rightly underline the hypocrisy of this dichotomy, according to which a technology is good if it can be used to produce insulin or select stem cells but is bad if used to produce seeds resistant to glyphosate. But the main danger of "ideological" use of the PP, in their opinion, lies in the fact that it distracts "politicians and consumers from true and significant threats to human health, and in so doing often prevents limited public health resources from being assigned to study them".[21] The EU allegedly "used and abused" the PP, ignoring all reports that clearly indicated that GM foods, obtained by recombinant DNA techniques, are merely

> [...] an extension, or refinement, of earlier, far less precise ones; adding genes to plants or microorganisms does not make them less safe either to the environment or to eat; the risks associated with recombinant DNA-modified organisms are the same in kind as those associated with conventionally modified organisms;

---

[18]S. Holm & J. Harris, 'Precautionary principle stifles discovery', *Nature* 400, 398 (1999).
[19]M. Peterson, 'The precautionary principle should not be used as a basis for decision-making', *EMBO Rep.* 8(4), 305–308 (2007).
[20]J. Hodgson, 'National politicians block GM progress', *Nat. Biotechnol.* 18, 918–919 (2000).
[21]H. I. Miller & G. Conko, 'Precaution without principle', *Nat. Biotechnol.* 19, 302–303 (2001).

and regulation should be based upon the risk-related characteristics of individual products, regardless of the techniques used
in their development.[22]

Application of the PP in Europe involved bureaucratic complications, administrative red tape, complex procedures to obtain the necessary authorizations and
extraordinary unsustainable protraction of development times,[23] for GMOs but not
for organisms obtained by traditional techniques. According to critics of the PP,
such procedures should only be required for the latter products. Indiscriminate,
acritical application of the PP may not only cause a sudden arrest of scientific
research but even violation of personal freedom "by ideologues who disapprove of
a technology on principle".

Miller and Conko's arguments can be accepted when they denounce a more
general irrational and anti-scientific danger fed by the resurgence of myths à la
Rousseau and the spread of New-Age-type occult pseudoculture that translates
into neo-Luddite movements animated by confused aversion to the modern world
and the idea of progress.[24] The distrust of biotech shown by large sectors of consumers cannot, however, be explained on the basis of emerging cultural fads
and trends. These are the result of a long series of economic and social disasters that have not only shown the limits of controls and the inadequacy of rules
designed to protect health but also irresponsible underestimation of risk and the
effects of not first rigorously evaluating technologies and production processes.
The list is long and includes the case of thalidomide, the recent scandal surrounding the drug Vioxx,[25] chicken factories à la dioxin and the epidemic of mad
cow disease.[26]

A number of uncertainties about the full effects of such foods
remain on the table (see Briefing, pages 651–656). In the case
of human health, these include potential allergenic reactions to
genetic changes that are not completely understood. As for the
environmental impacts, many scientists feel that widely quoted
"hazards", such as the potential spread of herbicide-resistant
"superweeds", have been overemphasized by critics. But there
is a broader consensus that the potential ecological disturbance
caused by a growing dependence on GM crops by modern farmers could be significant [...]. The failure to "prove" scientifically

[22] *Ibid.*
[23] S. L. Huttner, H. I. Miller & P. G. Lemaux, 'US agricultural biotechnology: status and prospects', *Technol Forecast. Social Change* 50, 25–39 (1995).
[24] W. J. Hanegraaff, *New Age Religion and Western Culture: Esotericism in the Mirror of Secular Thought*, State University, New York Press, New York (1998).
[25] R. Horton, 'Vioxx, the implosion of Merck and aftershocks at the FDA', *The Lancet* 364(9450), 1995–1996 (2004).
[26] For an instructive and disquieting review of food alterations and falsifications in Europe in recent decades, see the enquiry by J.-C. Jaillette, *Il cibo impazzito* (Feltrinelli, Milan (2001)) and F. Lawrence, *Non c'è sull'etichetta* (Einaudi, Torino (2005)).

that a new food is dangerous is not the same as proving it is safe — a lesson learnt from the BSE affair [...]. Industry complains that the public has lost trust in its scientific experts, but it will only make matters worse by declaring its own loss of trust in the judgement of the consumer. If labelling all foods produced by GM techniques, as many argue, turns out to be a necessary step in regaining trust on both sides, it could be a small price to pay.[27]

Opposition to GMOs has therefore seen the involvement of personalities with a wide range of cultural extractions and political persuasions, often animated by qualitatively different degrees and types of awareness: reducing it to extremist fringes of the anti-global movement is not only a gross simplification but an error. Thus when GMO proponents speak with insulting condescension, as they often do, it is hard not to obtain the impression that they are trying to avoid proper discussion and dodging specific questions. Criticism first arose in the scientific community and the doubts and concerns expressed are published in the same journals that host articles by enthusiasts of plant transgenesis. Besides the many papers documenting the utility and feasibility of experiments in genetic modification, the same journals publish critical articles by other exponents about their limits and risks. These are the people who need the answers and the answers must be scientific. The scarecrow of greens and no-globals will not fool anyone: the stage for discussion is in universities and research labs, not in the streets.

The recurrent attempt to delegitimize scientists who criticize GMOs is just as futile, bordering on pathetic. Scientists and researchers cannot be treated as ideological demagogues, their arguments distorted and simplified to the limit of caricature. It does not help bring out the arguments of either side, find solutions or promote real discussion.[28]

Despite the polemics, the PP and "subjective perception" as an element of risk have slowly begun to be assimilated as organizational principles of environmental jurisprudence, and above all to appear in international laws, albeit with some contradictions.[29] The behaviour of the WTO on two different occasions is indicative. On the first, the USA and Canada appealed to the WTO to invalidate the decision by Europe to ban imports of beef treated with hormones. Europe invoked the PP due to "possible" risks to health. The decision of the WTO Appellate Body confirmed

---

[27]'GM foods debate needs a recipe for restoring trust', *Nature* 398, 639 (1999).

[28]Examples include the statements reported in the volume edited by F. Battaglia and A. Rosati (*I Costi della Non-Scienza*, 21mo secolo, Milan, 2004) – a compendium of anti-environmental thought – in which paradoxical agreement of positions only apparently distant such as "scientific" Leninism and ultra-liberal capitalism, both characterized by faith in 19th century positivism can be found. It is incredible that such positions survive despite the advances of science.

[29]L. Busch, 'The homiletics of risk', *J. Agric. Environ. Ethics* 15, 17–29 (2002).

that the PP is not a set of normative rules or a legal principle and could not therefore be the basis of decisions that can only be justified on the basis of risk evaluation. The European decision was considered unfounded and "arbitrarily discriminatory", above all because the carcinogenic risk of meat treated with hormones was not significantly different from that of chickens raised with hormones, the meat of which was sold in European countries. The EU replied that scientific elements are not the only relevant factors, that the qualitative and not only quantitative dimension of risk had to be considered as well as "perception of risk", according to which Europeans considered beef with hormones to be much more dangerous than pigs with antibiotics. The WTO rejected "subjective perception" of risk as a constitutive element of risk, stating that application of the PP cannot ignore a "quantitative" evaluation of risk. Europe seemed to lose the appeal, until the WTO took the opposite position on another occasion involving France and Canada. Canada had appealed to annul France's decision to ban import of asbestos-like products, as dangerous for health. The WTO recognized that asbestos substitutes could not be discriminated as such (since the decision violated article III of GATT)[30] but nevertheless considered that the ban was well founded as "necessary to protect health". Subsequent appeals confirmed the first decision, making even more explicit the concept that threats to health are a factor of extraordinary importance in deciding whether or not a product is "similar" to one already banned (in this case asbestos). But

> [...] the Appellate Body went further. The Panel had refused to consider health risks as relevant to the "like products" analysis. The Appellate Body said this was wrong, and that health risk could be a very important factor in deciding whether products are "like." What is more, the Appellate Body said evidence of risk must be examined not only in the context of the physical properties of the products, but also in the context of consumer perception and behavior. This last point appears to open the door to evidence of risk perception, not just objective scientific evidence of risk [...]. Thus, at least in situations where the body of scientific evidence is inadequate to determine the objective health risks of a product, it could be argued that [...] the precautionary principle could fill the gap, at least for some WTO disputes. Even where subjective risk perceptions run counter to objective scientific evidence, those perceptions could be given weight in the "likeness" analysis, based on the language of the Appellate Body [...]. This would be especially true where a product has been banned or otherwise regulated based on consumer perception that the regulated product is riskier than unregulated products, even though risk assessment indicates that the risks are comparable [...]. Statements of the WTO Appellate Body carry great weight in future decisions. Accordingly, it can be anticipated that risk perception will play a major role in future deliberations of the WTO on issues such as genetically modified agricultural products.[16]

---

[30] General Agreement on Tariffs and Trade (GATT) (1994).

Scientific and Metascientific Criteria: The Nature of Risk

GMOs were created for different purposes. The first generation of GMOs can be classified in three groups according to the nature of the transgene inserted: GMOs expressing resistance to herbicide (thus enabling prolonged spraying without damaging the crop), to insects (expressing toxins specific for a predator) or to both (staked genes). These GMOs are the only ones to have reached the market and include soy, maize, rape and cotton. They were developed to improve company productivity, increase crop yields and reduce production costs. The second generation of GMOs includes plants modified to resist extreme environmental conditions (cold, drought) through reduced energy demand and longer conservation. The third generation regards GMOs with pharmacological characters, including anti-cancer plants, plants with increased vitamin content and edible vaccines. The technology enabling these goals cannot be considered a "simple" extension of "traditional" techniques, mainly based on grafting. Any claim to the contrary is an insult to people's intelligence and denigrates the extraordinary scientific progress underlying recombinant DNA technology. Varieties obtained traditionally by grafting experiments can only produce exchanges between species, if compatible, and are based on fusion of chromosomes that do not involve single genes. Besides, it would be mean to deny the absolute novelty of variants obtained with transgenic DNA, considering the extraordinary scientific commitment that went into their development. This novelty has also been formally recognized: a GMO is substantially different from its natural counterpart, since its "genetic material has been altered by procedures that do not occur naturally through mating and/or natural recombination".[31]

The position of GMO proponents on this *vaexata quaestio* is frankly paradoxical and contradictory. On one hand, GM foods are considered "substantially equivalent" to their natural counterpart, to the extent that they require few tests to declare their biocompatibility with ecological systems and human health. In this context, the technique of recombinant DNA is considered a simple(!) extension of "traditional" methods. On the other hand, GM foods are considered so "new" that they call for special laws to enable them to be patented and to defend intellectual property rights.

If we accept that the product is absolutely new, like the technology that created it, we need to clarify once and for all the nature of the risk that it can pose for human health and ecological equilibrium. It is now demonstrated that GMOs present a broad spectrum of dangers, like any human activity. The statement seems self-evident which may be why it is difficult to refute. Proponents of GMOs consider these risks to be largely predictable and in any case not quantitatively or qualitatively more severe than those presented by traditional techniques. However, this position is scientifically unsustainable, and it is regretful to have to point this out

---

[31] Council Directive no. 90/220/EEC.

to vehement upholders of evidence-based science. Contrary to the custom in other fields, as we have seen the risks of GMOs can neither be quantified nor predicted: this makes any attempt to define a reliable cost–benefit ratio somewhat elusive. As we know, drugs, the dangers of which are anything but potential, are authorized and prescribed on the basis of a good cost–benefit ratio and considered acceptable because of the expected benefit. In the case of GMOs, assessment of both costs and benefits is still hypothetical. Faced with such uncertainty, it is unrealistic to require well founded quantitative predictions. Nor can science be expected to express itself in the absence of certain data. Whether we like it or not, the decision about the acceptability of certain risks is not the exclusive task of scientists or even technocrats. This is particularly true in the light of two fundamental considerations. The first regards the objective nature of the risk, which in the case of uncontrolled spread of transgenic DNA would be largely irreversible, because it would threaten overall ecological equilibrium:

> The committee of 24 experts, both for and against GMOs, [...] recognized that risks for humans are very low [...] but the impact of fields of GMOs on biodiversity aroused concern. Three years of application of broad spectrum herbicides [made possible by cultivation of varieties engineered to resist them] led to a drastic drop in the number of weeds, as well as insects, small animals and birds. If this trend continues [...] biodiversity will be lost.[32]

Among the many "unexpected effects" that we have seen to characterize GMOs, this is one-way damage that cannot be "calculated".

The second consideration to bear in mind regards the metascientific nature of specific risk. Food is not a drug, nor an airbag, nor can it generally be likened to any other product of human activity. Beyond nutritional value, food embodies a set of composite values of a cultural, religious and symbolic nature that make it an absolutely unique good, the value of which cannot be calculated solely on the basis of economic or scientific parameters.[33]

Soylent Green

Many will remember the film *Soylent Green*, interpreted by Charlton Heston and directed by Richard Fleischer (1973).[34] The plot is set in "distant" 2022 when

---

[32] G. Romeo, 'Li piantiamo o non li piantiamo?', *Le Scienze* 425, 64–65 (2004) (Our translation).

[33] M. Megloughlin (Ed.), *Proceedings of the International Workshop on Animal Biotechnology issues*, The biotechnology program, Davis, CA (1994).

[34] The film (1973) was based on the novel by Harry Harrison *Make room! Make room!* published in 1966 in the magazine *Impulse*. The story was inspired directly by research commissioned by the *Club of Rome* (founded by Aurelio Pecci in 1960) conducted at the MIT and subsequently reported in D. H. Meadows, D. L. Meadows, J. Randers & W. W. Behrens III, *The Limits to Growth*, Universe Books, New York (1972).

food resources are based mainly on plankton after a series of catastrophic events. Actually, plankton has been depleted for some time and an unscrupulous biotech company, Soylent Green, produces biscuits allegedly made of soy (!) but actually made from human cadavers. The paradox at the heart of the drama lies in perception of the problem, not in the problem itself. Indeed, nobody disputes the fact that the composition of "green Soya" is "substantially equivalent" to that of normal food or that it meets human nutritional requirements, expressed in purely biochemical terms. The real question is another: what emerges is the process of alienation of the ethical and cultural dimension of food, as a value that does not reduce to "substance" – defined as "composition" in the most simplistic sense – or to market value and "goods". The concept of "substantial equivalence" removes the dimension of *context*, deliberately ignoring the method of production and the path by which food goes from producer to consumer, and denigrating qualitative aspects which are artificially limited to quantity calibrated on arbitrary parameters.

Paradoxically, this process is the complete opposite of that used by the EU to assign the "quality marks" Protected Designation of Origin (PDO) and Protected Geographical Indication (PGI)[35] for food produced in Europe. A GM food is by definition a food without any quality mark, since quality is not limited to composition (and we stress that knowledge of the "composition" of modified food can never be exhaustive). Food cannot be transformed by a "biochemical machine" – even if some retarded epigone of Descartes would like to reduce humans to a set of gears – but can be treated, cooked, prepared and consumed in the framework of a complex process that involves ritual, religious, historical, cultural, emotional, physiological and mental elements and reactions. To be acceptable, a food cannot be contrary to one's religious or philosophical beliefs,[36] it must be compatible with one's history and culture to be appetizing and digestible; it must meet aesthetic and psychophysical requirements. A good food is the product of a healthy science, a good tradition and much art. Daily experience is enough to demonstrate this: we would not look for refined and costly gourmet dishes, rare wines and special cheeses, if we were

---

[35]To protect typical products, the EU passed a law establishing two levels of recognition: PDO and PGI. The former extends protection of the national mark DOC (Denomination with Certified Origin) to all of Europe and through GATT to the rest of the world. PDO designates a product from a region or country having qualities or characteristics essentially or exclusively due to its geographical environment, a term that includes natural and man-made factors. The entire production and processing of the product must occur in the specific area. PGI introduces a new level of qualitative protection that considers industrial development in the sector, giving more weight to production methods than to geographic constraints. It indicates a product from a region or country, the quality, reputation and characteristics of which depend on geographical origin and of which at least one phase of production or processing occurred in the specific area.

[36]Inserting animal genes into food plants, especially genes from "unclean" (pig) or "sacred" (cow) animals, could cause tension or revolts among vegetarians and persons of religions that prohibit their use as food (Jews, Muslims, Hindus).

happy with a cheap, nourishing and chemically balanced, anonymous hamburger from a fast food bar. Indeed,

> although the vast majority of the European population is reluctant to accept GM food, many food products that contain GM ingredients are currently imposed on them with little possibility of choice and information. Because technical, objective scientific arguments are meant to be the only valid arguments, public suspicion of GMOs is interpreted as irrational and based on ignorance. But studies by the Economic and Social Research Council (ESRC, 1999) show that the public is not stupid and ignorant in its approach to risks, and that it has a qualified understanding of the main problems. The concept of substantial equivalence, introduced for the risk assessment of GM food, is a reducing concept because it ignores the context in which these products have been produced and brought to the consumer at the end of the food chain. Food quality cannot be restricted to mere substance and food acts on human beings not only at the level of nutrition but also through their relationship to environment and society [...]. Food products also mean processes. The content of food is more than its chemical composition and in the case of GM food, the public debate reaches far beyond food risk assessment. Behind substance, many other aspects are involved and can constitute or be perceived as risks or hazards for the consumer. The food chain process is the context. And the context can provide answers to questions such as who did what and how, when, and where? Context, the often forgotten factor when dealing with living processes, from genetics[37] (Holdrege, 1996) to human activities, needs to be re-introduced by considering these other aspects that are necessary to get from substantial equivalence to equivalence *per se*.[38]

It is embarrassing to list all the factors – place of origin, methods of production (use of pesticides, fertilizers, antibiotics) and transport – that are omitted when the criterion of substantial equivalence is used and which are nevertheless perceived as elements of risk evaluation by consumers. Engineering of foods violates product identity and

> directly influences one's health, physical constitution, and overall well-being. Consequently, food choice, be it reflective or unreflective, becomes a basic form of self-creating, self-expression, and self-definition. By making food choices, we literally create our bodies, our health, and ourselves, and affect directly the health of our children and loved ones for whom we care and with whom we share food. Therefore, to deprive a person of the ability to choose what goes into her body represents a violation of her authenticity and authorship in life that limits greatly that

---

[37] C. Holdrege, *Genetics and the Manipulation of Life: the Forgotten Factor of Context*, Lindisfarne Press, Hudson (1996).

[38] S. Pouteau, 'Beyond substantial equivalence: ethical equivalence', *J. Agric. Environ. Ethics* 13, 273–291 (2000).

person's ability to live according to her conception of the good life.[39]

New criteria and methods of assessment that consider this (rediscovered) dimension of risk are therefore needed, essentially an ethical and cultural dimension that restores consumer importance and dignity, making consumer perception and opinion an integral part of the risk assessment process,[40] not just a dependent variable that can be manipulated as the moment demands. Though media gurus completely ignore it,[41] discussion is raging not only in the scientific community[42,43] but also in the biotech industry that has begun to reflect on the ethical dimension of the problem.[44]

Absence of Proof Is Not Proof of Absence

Opponents of the PP sustain that there is no sure scientific data demonstrating the existence of significant risk and more generally testifying that GM foods are not safe. This statement contains a logical paradox that makes it unacceptable: absence of proof of risk does not demonstrate the inexistence of risk:

> However, at present the scientific information available is not sufficient to conduct reliable risk assessment for a proposed GMO release. Knowledge about probability and ecological effects of horizontal gene transfer from GMO to other organisms is lacking (3), and the inherent complexities and limitations in predictions of interactions and impacts on ecological systems may prohibit identification of important risks (4). In addition, present methods for the monitoring and detection of ecological effects are insufficient and this may result in inadequate control. Interpretations of risk may hence rely on commitment to specific objects or ambitions (71). Scientists working as counsellors in the regulatory process are in danger of being hijacked by those mainly interested in ensuring a specific regulatory outcome (72). Funding organizations might have different motiva tions for supporting research. Biotechnology companies, mostly

[39]A. Pascalev, 'You are what you eat: genetically modified foods, integrity and society', *J. Agric. Environ. Ethics* 16, 583–594 (2003).

[40]M. R. Schmidt & W. Wei, 'Loss of agro-biodiversity: uncertainty and perceived control: a comparative risk perception study in Austria and China', *Risk Anal.* 26(2), 455–470 (2006).

[41]J. Scholderer & L. J. Frewer, 'The biotechnology communication paradox: experimental evidence and the need for a new strategy', *J. Consumer Pol.* 26, 125–157 (2003).

[42]J. C. Polkinghorne, 'Ethical issues in biotechnology', *Tibtech* 18, 8–10 (2000).

[43]J. Kinderlerer, 'Is a European Convention on the ethical use of modern biotechnology needed?', *Trends Biotechnol.* 18, 87–90 (2000).

[44]R. Fears & E. Tambuyzer, 'Core ethical values for European bioindustries', *Nat. Biotechnol.* 17, 114–115 (1999).

> interested in economic growth, may form close connections with both scientists and policy regulators to favour a decision that promotes more favourable economic prospects [...]. Most scientists have problems with regard to critical and objective evaluation of their own work, particularly in the context of risk and harmful effects.[7]

The starting hypothesis of the experiments translates into an asymmetry that application of the PP attempts to balance by prevalently addressing "prevention of false negatives at the expense of a few false positives".

After hesitations and contrasts, risk analysis posed by GMOs obtained a first normative framework with adoption of the principle of substantial equivalence (PSE), which implies studying the composition of GM foods and comparing it with that of the natural counterparts. A "substantially equivalent" GM food is one that does not show significant differences with respect to its control, with which it shares the spectrum of risks, and can therefore be marketed. The reliability and intrinsic limits of this criterion have been amply discussed and criticized as insufficient, though it is an important advance with respect to the previous absolute absence of rules.

The PSE was drafted in 1993 and implemented and rewritten many times before its formulation of 2000.[45] Though ignored by those who consider the PSE to be the alpha and omega of GMO safety assessment, it is increasingly evident that its application does not guarantee and is not *per se* a certification of safety, though it enables potential differences to be identified; it is a starting point, not a destination.[46] Authoritative figures have therefore radically and rigorously criticized the PSE but have received no answer.

GMOs are compared with their non-modified counterpart, the safety of which is testified by custom. This is paradoxical, especially when upheld by those who rightly stress the dangers of certain "natural" foods and foods obtained by so-called organic techniques. In the last 20 years, alarming evidence of the presence of toxic or anti-nutritional substances in foods has emerged.[47] While precise risk identification criteria have been defined for chronic *exposure* to such foods, little

---

[45] FAO/WHO, Safety aspects of genetically modified foods of plant origin. Report of a joint FAO/WHO expert consultation on foods derived from biotechnology, Geneva, Switzerland 29 May–2 June 2000, FAO, Rome, available at: http://www.who.int/foodsafety/publications/biotech/en/ec_june2000_en.pdf.

[46] H. A. Kuiper, G. A. Kleter, H. P. J. M. Noteborn & E. J. Kok, 'Substantial equivalence – an appropriate paradigm for the safety assessment of genetically modified foods?', *Toxicology* 181–182, 427–431 (2002).

[47] A. J. Essers, G. M. Alink, G. J. A. Speijers, J. Alexander, P.-J. Bouwmeister, P. A. van der Brandt, S. Ciere, J. Gry, J. Herrman, H. A. Kuiper, E. Mortby, A. G. Renwick, D. H. Shrimpton, H. Vainio, L. Vittozzi & J. H. Koeman, 'Food plant toxicants and safety – risk assessment and regulation of inherent toxicants in plant foods', *Environ. Toxicol. Pharmacol.* 5, 155–172 (1998).

has been done to determine the effects and adverse reactions of chronic *consumption*. Such studies are complex and have unpredictable limits because foods are not only complex mixtures of thousands of biologically active compounds, but their composition also varies widely in relation to area of origin and methods of conservation and processing, so that "application of standard toxicity tests used to assess synthetic chemical molecules and food additives is virtually impossible".[46]

Comparison should therefore be made between the GM plant and the nearest direct or isogenic ancestral line. This is sometimes impossible, in which case many other lines must be considered to establish a comparison.[48] This should be done for a wide range of substances – macro- and micronutrients, toxins, anti-nutritional factors – including those depending specifically on expression of the transgene. Clearly this type of investigation requires rigorous standardization at international level, which despite the agreement reached for certain foods,[49] has not been done. However, procedures of risk determination substantially different from those with which the problem was tackled in the 1990s are needed: scientific research is dedicating increasing attention to this limit since classical assessments of safety and risk applied to toxic chemical substances and previous technologies are only partly applicable in assessment of the ecological and long-term effects of GMOs. GMOs do not have predictable relations between genotype and phenotype, similar to the relations between structure and activity of toxic substances that enable risk prediction.[50]

The chemical composition of normal and GM samples cultivated in different geographical areas and seasons should be analysed for unexpected effects on metabolic pathways leading to unintended metabolites or toxins. This aspect is of particular importance since an acceptable degree of difference in composition between modified and non-modified plants cannot be established, because natural varieties are already subject to wide intrinsic variability. The impossibility then of predicting unexpected effects – a phenomenon of which we have gradually become aware in the last 5 years – makes the concept of substantial equivalence completely aleatory.

Although at first impression, the approach of the PSE

> might seem plausible and attractively simple, we believe that
> it is misguided, and should be abandoned in favour of one

---

[48] OECD, *Safety Evaluation of Foods Derived by Modern Biotechnology: Concept and Principles*, Organization for Economic Cooperation and Development, Paris (1993), available at: http://dbtbiosafety.nic.in/guideline/OACD/Concepts_and_Principles_1993.pdf.

[49] OECD, *Consensus Documents, Organization for Economic Cooperation and Development*, Inter-Agency Network for Safety in Biotechnology, Paris (2001).

[50] O. Käppeli & L. Auberson, 'The science and the intricacy of environmental safety evaluations', *Tibtech* 15, 342–349 (1997).

that includes biological, toxicological and immunological tests rather than merely chemical ones [...]. The concept of substantial equivalence has never been properly defined; the degree of difference between a natural food and its GM alternative before its "substance" ceases to be acceptably "equivalent" is not defined anywhere, nor has an exact definition been agreed by legislators. It is exactly this vagueness that makes the concept useful to industry but unacceptable to the consumer.

Moreover, reliance of policymakers on the concept of substantial equivalence acts as a barrier to further research into the possible risks of eating GM foods.[51]

Millstone's article goes on to denounce the inadequacy of the PSE which he claims is not by chance: governments should have written laws that treated GM foods in the same way as "new chemical, pharmacological, pesticide and additive compounds", requiring companies to provide exact information enabling determination of an acceptable daily intake (ADI). These studies were not conducted because they would have cost too much ("about 25 million dollars per product"); this would have delayed market release of GMOs; above all, the definition of ADIs "would have restricted use of GM foods to a marginal role in the diet of consumers".[52] The companies presume that GMOs are safe – and ask governments to confirm it for worried consumers – but at the same time they ask governments to keep controls and testing to a minimum. The WHO and FAO have indeed followed this suggestion, requiring that modified foods be compared with the corresponding unmodified foods and calling for further tests, to decide case by case, only whether there are significant differences in chemical composition. According to Millstone, scientists are unfortunately not yet able to reliably predict the toxicological or biochemical consequences of a GMO from its composition. Despite its intrinsic ambiguity, the PSE is also applied in arbitrary and disputable ways. Millstone reports a concrete example concerning GM soy, a major battlehorse of the biotech companies:

GM glyphosate-tolerant soya beans (GTSBs) illustrate how the concept has been used in practice. The chemical composition of GTSBs is, of course, different from all antecedent varieties, otherwise they would not be patentable, and would not withstand the application of the herbicide glyphosate. It is quite straightforward to distinguish, in a laboratory, the particular biochemical characteristics that make them different. GTSBs have, nonetheless, been deemed to be substantially equivalent to non-GM soya beans, by assuming that the known

---

[51] E. Millstone, E. Brunner & S. Mayer, 'Beyond "substantial equivalence" ', *Nature* 401, 525–526 (1999).
[52] ADI is defined as 1% of the highest dose that has deleterious effects on the animal.

genetic and biochemical differences are toxicologically insignificant, and by focusing instead on a restricted set of compositional variables, such as the amounts of protein, carbohydrate,
vitamins and minerals, amino acids, fatty acids, fibre, ash,
isoflavones and lecithins. GTSBs have been deemed to be
substantially equivalent because sufficient similarities appear
for those selected variables. But this judgement is unreliable.
Although we have known for about ten years that the application
of glyphosate to soya beans significantly changes their chemical composition (for example, the level of phenolic compounds
such as isoflavones[53]) the GTSBs on which the compositional
tests were conducted were grown without the application of
glyphosate![54] This is despite the fact that commercial GTSB
crops would always be treated with glyphosate to destroy surrounding weeds. The beans that were tested were, therefore,
of a type that would never be consumed, while those that are
being consumed were not evaluated. If the GTSBs had been
treated with glyphosate before their composition was analysed,
it would have been harder to sustain their claim to substantial
equivalence.[50]

Obviously one can argue about the importance of changes in the biochemistry
of plants introduced by transgenesis. They could be desirable, toxic or neutral.
However, the fact remains that this question cannot be answered because an answer
was not sought in the appropriate manner. This may be the main limitation of the
PSE: faith in its reliability actually deceives those with government responsibility,
blocking further research to reveal risks connected with GMOs. The PSE

[...] is a pseudo-scientific concept because it is a commercial
and political judgement masquerading as if it were scientific.
It is, moreover, inherently anti-scientific because it was created
primarily to provide an excuse for not requiring biochemical
or toxicological tests. It therefore serves to discourage and
inhibit potentially informative scientific research [...]. If policymakers are to provide consumers with adequate protection,
and genuinely to reassure them, then the concept of substantial equivalence will need to be abandoned, rather than merely
supplemented. It should be replaced with a practical approach
that would actively investigate the safety and toxicity of GM
foods rather than merely taking them for granted, and which
could give due consideration to public-health principles as well
as to industrial interests.[50]

---

[53] J. J. Lydon & S. O. Duke, 'Pesticide effects on secondary metabolism of higher plants',
Pest. Sci. 25, 361–374 (1989).
[54] S. R. Padgette, N. B. Taylor, D. L. Nida, M. R. Bailey, J. MacDonald, L. R. Holden &
R. L. Fuchs, 'The composition of glyphosate-tolerant soybean seeds is equivalent to that
of conventional soybeans', *J. Nutr.* 126, 702–716 (1996).

Millstone's criticism, as expected, aroused strong negative reactions[55,56,57] as well as much agreement,[58,59,60,61] demonstrating inexistence of the unanimity claimed by self-appointed representatives of "Science" who only award patents and prizes to those who share their faith in biotech.[62] According to Millstone and collaborators, certain enthusiasts of GMOs seem unwilling to face the scientific challenge and legislative problems that have been raised.[63] Science and scientists who consider themselves advocates of science would probably cut a better figure and reacquire lost trust by stating that

> [...] uncertainty can be explicitly stated and reduced by repro-
> ducible experiments under controlled conditions. However, the
> domain of ignorance, characterized by the interaction between
> unknown processes and/or unknown state-variables, tends to
> be implicitly neglected in risk assessment. As the well-known
> examples of dichlorodiphenyltrichloroethane (DDT) and chloro-
> fluorocarbons (CFCs) suggest, our ability to assess novel risks
> is primarily limited by the fundamental difficulty of taking
> these interactions into account [...]. To overcome mutual mis-
> understanding by scientists, policymakers and the public, it is
> important for all to acknowledge that unanticipated effects of

[55]A. Trewavas & C. J. Leaver, 'Conventional crops are the test of GM prejudice', *Nature* 401, 640 (1999).

[56]M. J. Gasson, 'Genetically modified foods face rigorous safety evaluation', *Nature* 402, 229 (1999).

[57]D. Burke, 'No GM conspiracy', *Nature* 401, 640–641 (1999).

[58]S. L. Taylor & S. L. Hefle, 'Seeking clarity in the debate over the safety of GM foods', *Nature* 402, 575 (1999).

[59]M. Tester, 'Seeking clarity in the debate over the safety of GM foods', *Nature* 402, 575 (1999).

[60]M.-W. Ho, 'Seeking clarity in the debate over the safety of GM foods', *Nature* 402, 575 (1999).

[61]A. Aumaitre, K. Aulrich, A. Chesson, G. Flachwosky & G. Piva, 'New feeds from geneti-cally modified plants: substantial equivalence, nutritional equivalence, digestibility and safety for animals and the food chain', *Livestock Prod. Sci.* 74, 223–238 (2002).

[62]These positions, distinguished by the smug, arrogant tones with which they avoid rather than tackle the problems raised, are summed up by the following passage: *Scientists who work in research laboratories denounce the dangers of non-science ... Regarding the long-term effects of GM plants ... the problem has been debated in the scientific community for years and ... the conclusion is that the probability of long-term effects, different from those associated with conventional plants, cannot be sustained.* This implies that those with different positions are not real scientists because they boast non-existent publications and university titles [...] and have no experience even though they sit on technical ministerial commissions (F. Sala, *Gli OGM sono pericolosi?*, Laterza, Bari (2005), pp. 36, 91–92). Such accusations could easily be aimed at sustainers of GMOs, among whose ranks there are of course respectable scientists, but usually persons who have nothing to do with biology or agriculture. In any case, current specific questions of interest are not mentioned and no attempt is made to answer them. Millstone's article is not even cited.

[63]E. Millstone, E. Brunner & S. Mayer, 'Seeking clarity in the debate over the safety of GM foods (reply)', *Nature* 402, 575 (1999).

novel technologies are not just possible but probable – and
that potential harmful consequences cannot reliably be established by further research since they fall into the domain of
ignorance. Risk assessment and policy need to emphasize the
limits of knowledge, rather than proving existing knowledge to
be correct.[64]

Data published very recently confirms these observations and helps to underline
that the scientific uncertainty inherent in risk assessment of new technologies calls
for a new study method. Thus it has finally emerged that a correct interpretation
of the PP translates into a significant increase in scientific research,[65] contrary to
what opponents say, whereas biased results and disgraceful examples of "omitted
research",[66] springing from prejudice and conflicts of interest of large sectors of
the academic world, have made it impelling to start independent research.

Declarations of faith in the criterion of substantial equivalence have not convinced most scientists or government authorities, who have finally expressed[67] the
need for a specific risk assessment process based on an integrated holistic approach,
aimed at recording expected and unexpected repercussions on human health and
on natural and agricultural ecosystems.

Since the early 2000s, many researchers, agreeing with Millstone's words, have
taken a critical look at the criterion of substantial equivalence, suggesting new and
more articulated study protocols with perspectives that differ radically from that
inspiring the PSE, and which would offer major guarantees if applied correctly on
a routine basis.[68,69] Most of these recommendations have been adopted by the
European authorities, which have more or less explicitly considered the PSE
"unsuitable for regulatory decision making"[70] since 2001, and have suggested a
complete validation procedure, aimed among other things at systematic detection
of "unexpected effects", for every new GMO.[71]

[64]H. Hoffmann-Riem & B. Wynne, 'In risk assessment one has to admit ignorance', *Nature*
416, 123 (2002).
[65]C. Raffensperger, J. Tickner, T. Schettler & A. Jordan, ' "... and can mean saying 'yes'
to innovation" ', *Nature* 401, 207–208 (1999).
[66]S. Garattini, 'The risk of bias from omitted research', *Br. J. Med.* 321, 845–846 (2000).
[67]WHO, *Study on Modern Food Biotechnology, Human Health and Development* (2005),
available at: http://www.who.int/foodsafety/biotech/who_study/en/index.html.
[68]P. Schenkelaars, 'Rethinking substantial equivalence', *Nat. Biotechnol.* 20, 119 (2002).
[69]Royal Society of Canada, *Genetically Modified Plants for Food Use and Human
health – An Update*, Royal Society of Canada, Ottawa, ON, Canada (2002).
[70]P. Schenkelaars, *op. cit.*
[71]P. Schenkelaars, Biotechnology Consultancy *GM Food crops and application of
substantial equivalence in the European Union.* Commissioned by Consumentenbond
and the Dutch Foundation Consument & Biotechnologie, SBC Leiden, the Netherlands
(June 2001), available at: http://www.consubiotech.nl/.

The Unpredictability of Risk

When a GMO is created, certain characteristics of the organism may be altered unintentionally. The transgenic DNA segment may be inserted in a pre-existing gene or nearby, upsetting its expression and function. The transgene may be activated too soon by promoter segments or after interaction of open reading frame (ORF) segments of the host with the transgenic DNA promoter. Rearrangement of transgenic DNA during insertion in the host genome may form spurious ORFs which in turn can induce synthesis of unintended substances. Recombination after introduction of repeated sequences in the transgene can cause chromosome instability at the insertion locus or ectopic recombination phenomena. Finally, hyper- or hypomethylation due to the transgenesis operation can silence one or more genes and/or interfere with the expression of other DNA segments, inducing complex pleiotropic effects. Finally, interaction of the transgene with environmental factors, especially stress, can give rise to a wide variations in biological behaviour and manifestation of the properties of the modified plant, producing a spectrum of unintended effects. Some of these unintended consequences can be partly predicted (if the exact site of DNA insertion, the function of the transduced segments and their involvement in host metabolic processes are known), "while others are not predictable, essentially because we do not know the regulation of the gene and its interaction with other DNA segments". Presumably certain of these effects may never manifest or only express under certain culture or GMO treatment conditions. This intrinsic uncertainty of GMOs, that has long been reported, is finally recognized:

> A WHO/FAO expert consultation in 2003 acknowledged that introduction of a transgene into a recipient organism is not yet a precisely controlled process and can result in a variety of outcomes regarding integration, expression, and stability of the transgene in the host[72] [...]. The consultation concluded that a consideration of compositional changes is not the sole basis for determination of safety and that safety can be determined only when the results of all aspects under comparison, and not merely comparisons of key constituents, are integrated.[73]

Unintended effects due to chromosome rearrangements or altered genome function have actually also been detected in natural foods produced by traditional grafting or subject to irradiation or chemical treatment to induce mutations that could

---

[72]FAO/WHO, *Safety Assessment of Foods Derived from Genetically Modified Animals including Fish*. Joint FAO/WHO expert consultation on food derived from biotechnology (2003), available at: http://www.who.int/foodsafety/biotech/meetings/en/gmanimal_reportnov03_en.pdf.
[73]A. G. Haslberger, 'Need for an integrated safety assessment of GMOs linking food safety and environmental considerations', *J. Agric. Food Chem.* 54, 3173–3180 (2006).

be "useful".[74,75,76] Processed foods polluted by pesticides or other chemical substances or affected by environmental degradation contribute to an increasing number of diseases and are a major public health problem.[77] This is an aspect that ingenuous greenies forget. Even modifications by traditional techniques should undergo rigorous assessment. However, the frequency of unintended effects in traditional agricultural production, where efficient selection of lines with desirable qualities and rejection of the others has been carried out for a long time, is much lower[78] than that observed in GMOs: In cultures based on traditional methods and mutagenesis, the selection process eliminates accidental variants that could be dangerous. Regarding GMOs, the question is whether unexpected or extra risks may not be detected by the processes of screening.[79] With recombinant DNA methods, very few plants in a GM population behave in the expected way,[80] which implies the need for adequate selection of specimens that meet health and productivity criteria. Genetic modification of a seed is not the same as creating plant hybrids by traditional methods. It is surprising that people deny this!

> Although the biological world works by variation and selection, this is generally accomplished in the context of a normal complement of endogenous genes that, though perhaps different, are allelic. This is quite distinct from GM plants where many copies of a gene are introduced and integrated randomly [...] genetic modification is also distinct from breeding two strains that have been safely consumed for extended periods of time.[81]

GM plants have an extra risk factor due to introduction of a foreign gene and effects on transgene expression mediated by interaction of environmental and

---

[74]H. A. Kuiper, G. A. Kleter, H. P. J. M. Noteborn, E. J. Kok, 'Assessment of food safety issues related to genetically modified foods', *Plant J.* 27(6), 503–528 (2001).

[75]T. Lincoln, 'Danger in the diet', *Nature* 448, 148 (2007).

[76]A. Konig, A. Cockburn, R. Crevel, R. W. Debruyne, R. Graftstoem, U. Hammerling, I. Kimber, I. Knudsen, H. A. Kuiper & A. A. Peijnenburg, 'Assessment of the safety of foods derived from genetically modified (GM) crops', *Food Chem. Toxicol.* 42, 1047–1088 (2004).

[77]O. Fennema, 'Influence of food–environment interactions on health in the twenty-first century', *Environ. Health Perspect.* 86, 229–232 (1990).

[78]F. Cellini, A. Chesson, I. Colquhoun, A. Constable, H. V. Davies, K. H. Engel, A. M. R. Gatehouse, S. Karenlampi, E. J. Kok, J.-J. Leguay, S. Lehesranta, H. P. J. M. Noteborn, J. Pedersen & M. Smith, 'Unintended effects and their detection in genetically modified crops', *Food Chem. Toxicol.* 42, 1089–1125 (2004).

[79]O. Käppeli & L. Auberson, 'How safe is safe enough in plant genetic engineering?', *Trends Plant Sci.* 3(7), 276–281 (1998).

[80]M. De Block, 'The cell biology of plant transformation: current state, problems, prospects and the implications for plant breeding', *Euphytica* 71, 1–14 (1993).

[81]D. Schubert, 'A different perspective on GM food (reply)', *Nat. Biotechnol.* 20, 1197 (2002). Schubert goes on to say that "Parrot and colleagues cite multiple gene deletion in variety of maize to illustrate their argument that 'unintentional consequences' are much more likely to occur in breeding than in biotechnology. However maize is unique because its genome comprises 80% retrotransposons; it has evolved to deal with redundancy; must other species have not."

genetic factors.[82] Variations in environmental conditions probably influence the frequency and characteristics of unintended effects expressed by GMOs more than genomic heritage,[83] and far from being reassuring, this finding testifies that processes underlying the expression of unintended effects are even less controllable and predictable:

> Hence the very process of genetic engineering has a unique capacity to produce harmful effects, and contrary to the recommendations of the *National Research Council* report, this justifies the opinion that all varieties produced by recombinant DNA should be examined with particular attention and checked for these effects. In view of the unknown nature of the danger, it is not clear how exactly to proceed. Even wide monitoring of a large variety of animals does not provide any guarantee of safety because certain plants are toxic for some species and not for others. So far there have been no known cases of poisoning by transgenic plants. A few times, however, we went close![84]

The writer of the aforementioned passage is not a complaisant journalist or a militant greenie, but Richard Lewontin, director of research at Harvard University and genetist of world renown. The existence of unintended effects and their unpredictability has made it necessary to reconsider the overall approach to risk assessment used until a few years ago. In 2003, the Ad Hoc Intergovernmental Task Force on Food Derived from Biotechnology of the Codex Alimentarius Commission established new assessment procedures that clearly favour a solution in continuity with the past and open new horizons.[85] The impossibility of defining general rules valid for all GMOs makes it necessary, though not sufficient, to conduct exhaustive molecular characterization of every genetic transformation (including analysis of the construct, its insertion site and the presence of any truncated segments) and its possible unexpected effects, evaluating case by case, when and whether this strategy can or must be improved or widened. However, to ascertain ecological impacts and effects on human health, more is needed. The possible consequences of genetic contamination and its effects on biodiversity must be properly studied in the field over long periods;[86] health implications, on the other hand, call for epidemiological

---

[82]A. G. Haslberger, 'Codex guidelines for GM foods include the analysis of unintended effects', *Nat. Biotechnol.* 21, 739–741 (2003).

[83]W. Novak & A.G. Haslberger, 'Substantial equivalence of antinutrients and inherent plant toxins in genetically modified novel foods', *Food Chem. Toxicol.* 38, 473–483 (2000).

[84]R. Lewontin, *Il sogno del Genoma Umano ed altre illusioni della Scienza*, Laterza, Bari (2002), p. 280 (our translation).

[85]Codex Alimentarius Commission, Joint FAO/WHO Food Standard Programme. *Codex Ad Hoc Intergovernmental Task Force on Food Derived from Biotechnology* (Codex Yokohama Japan), available at: http://www.who.int/fsf/GMfood/codex_index.htm.

[86]L. Firbank, M. Lonsdale & G. Poppy, 'Reassessing the environmental risks of GM crops', *Nat. Biotechnol.* 23(12), 1475–1476 (2005).

controls and programmes of post-market surveillance on selected samples of population over significant time intervals.[87] Companies producing GMOs now seem to realize the complexity of the problem and recognize that very little experience has been acquired in the assessment of the safety of GM foods not substantially equivalent to their counterparts; however, research is underway to develop satisfactory protocols for these products.[88]

The last 5 years have seen a burgeoning of new research inspired by the integrated approach of systems biology and based fundamentally on new high-throughput techniques with multivariate analysis, such as proteomics, functional genomics and especially metabolomics.[89,90,91] The main aim of evaluating the safety of GM foods is to identify unintended effects and unexpected modifications due to pleiotropic transformations caused by the transgene. Metabolomics that provides an analytical and quantitative description of a vast spectrum of substances and biochemical pathways can answer these questions, though not exhaustively. For it to become the reference method, a database of control and modified products must be set up to enable useful and significant comparisons. My dear friend, Filippo Conti, professor of Physical Chemistry at Rome University La Sapienza, pioneer of the new method and president of the Italian Association of Metabolomics, believes that this holistic approach is the way to obtain a convincing dynamic assessment of risks and characteristics of GMOs:[92] traditional analysis can offer little certainty,[93] as some researchers of biotech companies are reluctantly admitting.[94] Unfortunately, these new applications are exceptions, limited to a few universities, and they do not consider additional factors of risk assessment – potential contamination of

---

[87] J. J. Hlywka, J. E. Reid & I. C. Munro, 'The use of consumption data to assess exposure to biotechnology-derived foods and the feasibility of identifying effects on human health through post-market monitoring', *Food Chem. Toxicol.* 41, 1273–1282 (2003).

[88] K. T. Atherton, 'Safety assessment of genetically modified crops', *Toxicology* 181–182, 421–426 (2002). It is surprising that these considerations are made by researchers working for GM-producing companies, while quite different independent scientists and journalists say the opposite (for an edifying review of their positions, see: F. Sala, *Gli OGM sono pericolosi?*, Laterza, Bari (2005); A. Meldolesi, *Organismi geneticamente modificati: storia di un dibattito truccato*, Einaudi, Torino (2001)).

[89] H. A. Kuiper, E. J. Kok & K.-H. Engel, 'Exploitation of molecular profiling techniques for GM food safety assessment', *Curr. Opin. Biotechnol.* 14, 238–243.

[90] I. J. Colquhoun, G. Le Gall, K. A. Elliott, F. A. Mellon & A. J. Michael, 'Shall I compare thee to a GM potato?', *Trends Genet.* 22(10), 525–528 (2006).

[91] H. Rischer & K.-M. Oksman-Caldentey, 'Unintended effects in genetically modified crops: revealed by metabolomics?', *Trends Biotechnol.* 24(3), 102–104 (2006).

[92] C. Manetti, C. Bianchetti, M. Bizzarri, L. Casciani, C. Castro, G. D'Ascenzo, M. Delfini, M. E. Di Cocco, A. Laganà, A. Miccheli, M. Motto & F. Conti, 'NMR-based metabonomic study of transgenic maize', *Phytochemistry* 65, 3187–3198 (2004).

[93] J. A. Heinemann, A. D. Sparrow & T. Traavik, 'Is confidence in the monitoring of GE foods justified?', *Trends Biotechnol.* 22(7), 330–336 (2004).

[94] A. Cockburn, 'Assuring the safety of genetically modified (GM) foods: the importance of an holistic, integrative approach', *J. Biotechnol.* 898, 79–106 (2002).

ecosystems, loss of acquired genetic characters, spread of herbicide resistance –
that are of extraordinary importance in evaluating not only ecological impact but
also medium- and long-term utility and economic benefits.[95]

Despite all these precautions and the complexity of measures devised to tackle
the possibility of unexpected negative consequences, the risk associated with GMOs
is still uncertain:

> Hazards due to genetic manipulation and, more generally, to
> biological entities, are not as explicit and easy to identify as,
> for instance, hazards due to the use of chemical pesticides. For
> instance, a GMO released for weed control might escape and
> become pathogenic to cultivated crops or native plants: the
> hazard is identified but its timing, probability, extent, and con-
> sequences are still impossible to predict. Scientific methods in
> systematic hazard identification are not yet available for micro-
> organisms [...]. No list can possibly contain unexpected hazards.
> Similarly, the determination of probability (likelihood) of putative
> hazards posed by GMOs, i.e. the if and when a given event will
> occur, relies on simple conjectures.[96]

This is not a partisan opinion, since it is also shared by strenuous proponents
of GMOs: "since we have no data to evaluate exposure and dose-response relationships, we cannot appropriately characterize risk as understood by the scientific
community".[97,98] High gene instability and genomic variability of modified species
observed in different lots of seeds not only confirm the variability in behaviour and
characteristics of the new varieties but also make it impossible to evaluate once
and for all their ecological impact and effects on health: with every experiment
the product is "different" and all the tests have to be done again from scratch!

Besides applying these new analytical techniques, it is now clear that the introduction of GM foods, like drugs, must be subject to a rigorous system of monitoring
after release (post-market surveillance) in order to evaluate their medium- and
long-term effects. Such studies are the gold standard for identifying unintended
and unpredictable effects, as in the pharmacological sector. The analogy with phar-
macology, however, ends here, because in the sector of agriculture and food, there
is no reference figure like doctors who play a key role in the drug surveillance

---

[95] M. P. K. Von Krauss, E. A. Casman & M. J. Small, 'Elicitation of expert judgments of
uncertainty in the risk-assessment of herbicide-tolerant oilseed crops', Risk Anal. 4(6),
1515–1527 (2004).
[96] Q. Migheli, 'Genetically modified biocontrol agents: environmental impact and risk
analysis', J. Plant Pathol. 83(2), 47–56 (2001).
[97] E. Hodgson, 'Genetically modified plants and human health risks: can additional
research reduce uncertainties and increase public confidence?', Toxicol. Sci. 63, 153–156
(2001).
[98] A. Bakshi, 'Potential health effects of genetically modified crops', J. Toxicol. Environ.
Health B Crit. Rev. 6(3), 211–225 (2003).

service. As well, the drugs are completely separate from any other product taken by patients and are used for definite, limited periods. This means that side-effects are much more easily identified for drugs than for food. Different methods of control have been devised for food,[99] and all require as a preliminary condition clear labelling and traceability of the product, so that the production line from producer to consumer can be easily reconstructed.[100] Unfortunately these strategies cannot be applied to GMOs in the USA because labelling is not required or is incomplete, whereas in Europe percentages of GM foods below 0.9% do not have to appear in the label. "Only acute effects associated with elevated consumption of a substance can probably be detected by post-market surveillance; infrequent or long-term effects generally call for specific epidemiological techniques, application of which requires much more data than is currently collected. We therefore know very little about the potential long-term effects of any food."[101]

Moral: the EU has required post-market surveillance since 1997[102] and the directive was ratified by the FAO/WHO in 2000;[103] nevertheless, no rigorous study on the long-term impact of GM foods on health is yet available and probably cannot be expected even in the medium period. The reason is probably that such investigations are expensive (companies have no intention of paying for them) and first require labelling of the product and farm-to-consumer traceability.[104]

On the latter aspect, something seems to be changing slowly. To enable longitudinal post-market surveillance, to segregate modified from natural seeds, to

---

[99] J. J. Hlywka, J. E. Reid & J. C. Munro, 'The use of consumption data to assess exposure to biotechnology-derived foods and the feasibility of identifying effects on human health through post-market monitoring', *Food Chem. Toxicol.* 41, 1273–1282 (2003).

[100] An interesting experiment was conducted by the association *Un Punto Macrobiotico* (UPM), presided by the Italian Mario Pianesi. UPM only sells food produced and managed in a traditional manner – without GM varieties or treatment with pesticides – verifying origin and quality of seeds, soil and irrigation systems. The label gives full details of composition and processing. Retrospective checks of organoleptic quality and unexpected effects are conducted thanks to information collected at UPM restaurants and from consumers, which together with a scientific committee of international experts, ensures an effective supervision network on the whole production and supply line of the goods. The model developed by Pianesi was discussed and considered by MIPAF.

[101] H. A. Kuiper, G. A. Kleter, H. P. J. M. Noteborn & E. J. Kok, 'Assessment of food safety issues related to genetically modified foods', *Plant J.* 27(6), 503–528 (2001).

[102] EU 97/618/EC: Commission Recommendation of 29 July 1997 concerning the scientific aspects and the presentation of information necessary to support applications for the placing on the market of novel foods and novel food ingredients and the preparation of initial assessment reports under Regulation (EC) No. 258/97 of the European Parliament and of the Council, *Off. J. Eur. Commun.* L253, 1–36, available at: http://eur-lex.europa.eu/LexUriServ/LexUriServ.do?uri=CELEX:31997R0258:EN:HTML

[103] FAO/WHO, *op. cit.*

[104] M. Miraglia, K. G. Berdal, C. Brera, P. Corbisier, A. Holst-Jensen, E. J. Kok, H. J. P. Marvin, H. Schimmel, J. Rentsch, J. P. P. F. van Rie & J. Zagon, 'Detection and traceability of genetically modified organisms in the food production chain', *Food Chem. Toxicol.* 42, 1157–1180 (2004).

inform citizens about what they eat (so that they can avoid foods that might cause allergy or intolerance), there have been many requests that marketed GM foods be accompanied by a labelling schedule, possibly validated by an independent body, as requested in the USA,[105] in order to ascertain the presence of GM components *irrespective of legal tolerance thresholds*. The EU made this obligatory and countries such as Australia, Japan, Greece and New Zealand have passed or are passing similar systems of traceability.[106] Despite increasing demand for labelling by citizens' associations, scientists and even control bodies, the USA has always been opposed to such measures on the pretext that there are no recognized differences between natural and GM varieties. An unpleasant consequence of this has been that GM seeds are disposed of, transported, stored and treated in the same facilities as normal seeds, leading to mixing which has caused many cases of contamination. This situation cannot continue with the globalization of markets and the spread of protests. Many nations require exhaustive certifications and rigorous segregation of different varieties; the biotech industry was forced to begin to tackle the problem and soon realized the extra costs involved.[107] If the system of labelling and identity preservation of commercial varieties becomes general, as seems inevitable, the costs will probably outweigh the advantages.[108] The biotech industry is unlikely to have thought of this and it will become a serious mortgage on the introduction of second and third generation varieties.[109]

Who Controls Who?

During the controversy on transgenesis in early 2000, an article in *Science*[110] asked, "If GM food is safe, where's the evidence?" The question cast doubt on the arguments of GM producers and their experts, and made fun of the alleged scientific nature of "evidence-based medicine", according to which procedures cannot be authorized unless validated by rigorous experimentation.

Exponents of this "new medicine" were also the most intransigent defenders of GM food. Curiously, there is almost no evidence about the safety of GM food. In a review, Prof. J. Domingo, author of the article in *Science*, noted a great lack

---

[105] W. E. Huffman, 'Production, identity preservation and labeling in a marketplace with genetically modified and non-genetically modified foods', *Plant Physiol.* 134, 3–10 (2004).
[106] T. H. Varzakas, G. Chryssochoidis & D. Argyropoulos, 'Approaches in the risk assessment of genetically modified foods by the Hellenic Food Safety Authority', *Food Chem. Toxicol.* 45, 530–542 (2007).
[107] Anon, 'IP crops a niche market now, but ...', *Inform* 12, 840–843 (2001).
[108] M. Rousu, W. E. Huffman, J. F. Shogren & A. Tegene, 'Effects and value of verifiable information in a controversial market: evidence from lab auctions of genetically modified foods?', *Rev. Agric. Econ.* 12, 34–46 (2003).
[109] T. McKeon, 'Genetically modified crops for industrial products and processes and their effects on human health', *Trends Food Sci. Technol.* 14, 229–241 (2003).
[110] J. Domingo, 'If GM food is safe, where's the evidence', *Science* 288, 1748 (2000).

of scientific literature on the safety of transgenic foods. Regarding their toxicity, he found only 45 references, only one of which regarded an experimental study. Seven others were letters to editors of scientific journals, comments, attitudes or simply opinions without any appreciable experimental result. Seven more were only partly relevant to the question. A second search, aimed at finding articles of "unintended effects" of transgenic foods, produced 67 references, only one of which was really pertinent to GMOs. The third search, directed generically to "GM food", produced 101 references, only four of which regarded experiments aimed at discovering effects on health. Finally, a search of the Toxline database, a sort of "yellow pages" of known toxins, did not mention a single compound related to GMOs, except for a generic mention of certain possible allergens found in food. On the basis of these results, Domingo ironically proposed two explanations: GM foods are probably safe, though the companies producing it declined to publish the data that could confirm this ... in which case, the evidence was implicitly assumed! Alternatively, it was more likely that specific studies to determine the health effects of transgenic foods had never been done. The Spanish scientist therefore concluded that if the former hypothesis were true, the companies should be forced to publish their data so it could be examined and validated; if the latter hypothesis were true, GM foods clearly could not be marketed until all the necessary research had been done. Useless to say, this grave and elegantly formulated accusation has not yet received a single reply.

Ten years later, the picture has not change significantly. As stated by Domingo, "The number of citations found in databases (PubMed and Scopus) has dramatically increased since 2006. However, new information on products such as potatoes, cucumbers, peas or tomatoes, among others was not available [...] [indeed] the number of studies specifically focused on safety assessment of GM plants is still limited."[111]

In addition, it should be noted that most of these studies have been conducted (only) by the biotechnology companies responsible for marketing these GM plants. Considering the high financial stakes involved, there is increasing concern that articles reporting health risks or nutritional value of GM food products, published in peer-reviewed journals, may be influenced conflicts of interest. In a study[112] involving 94 articles selected by objective criteria, financial or professional conflict of interest was found to be associated with study outcomes that cast GM products in a favourable light ($p = 0.005$). While financial conflict of interest alone did not correlate with research results ($p = 0.631$), a strong association was found between

[111] J. L. Domingo & J. G. Bordonaba, 'A literature review on the safety assessment of genetically modified plants', *Environ. Int.* 37, 734–742 (2011).
[112] J. Diels, M. Cunha, C. Manaia, B. Sabugosa-Madeira & M. Silva, 'Association of financial or professional conflict of interest to research outcomes on health risks or nutritional assessment studies of genetically modified products', *Food Policy* 36, 197–203 (2011).

author affiliation with industry (professional conflict of interest) and study outcome ($p < 0.001$). Can we trust them?

Equivalent for Whom?

More than 2000 new transgenic constructs have been tested in the field,[113] more or less everywhere, mostly without proper monitoring or observance of rules: very little research has been dedicated to interactions and the impact of this release on natural biota.[114] Indeed, more time is spent imagining all the possible benefits than evaluating possible dangers.[115,116] Risk evaluation is thus inadequate[117] and when exhibited, is usually based on documentation provided by the producer. Though rigorous on other questions, the FDA considers these certifications to be satisfactory and does not require studies by independent groups. GM products are considered safe until proven otherwise. Thus, when the FDA announced in 1992 that special labelling for genetically manipulated foods would not be necessary, the decision was hotly contested. The editorial board of the *New England Journal of Medicine* was critical, observing that the FDA's position seemed to favour industry rather than protect consumers.[118]

As pointed out by the *British Medical Journal*, the concept of health entertained by the FDA and biotech companies is somewhat limited:

> The premise on which most modern medical science has been based is that health is the product of the absence of disease and can be achieved through use of a vast array of drugs and, more recently, medicines derived from genetic engineering. The opposing view is that health is not merely the absence of disease but is in fact a dynamic equilibrium, which occurs when an organism is in a harmonious balance with its environment. Healthy plants, animals, and people are the result of sound husbandry and management, not the product of prophylactic doses of pharmaceuticals and genetically engineered drugs. The medical world would do well to entertain these types of ideas more seriously.[119]

---

[113] OECD, Group of National Experts of Safety in Biotechnology. *Analysis of Field Release Experiments*, 14 May, OECD, Paris (1993).

[114] S. Krimsky, *Biotechnics and Society. The Rise of Industrial Genetics*, Praeger, New York (1991).

[115] J. Dekker & G. Comstock, Ethical and environmental considerations in the release of herbicide resistant crops', *Agric. Human Values* 9(3), 31–43 (1992).

[116] B. M. Chassy, 'Food safety risks and consumer health', *New Biotechnol.* 27, 535–544 (2010).

[117] M. G. Paoletti & D. Pimentel, 'The environmental and economic costs of herbicide resistance and host–plant resistance to plant pathogens and insects', *Tech. Forecast. Social Change* 50, 9–23 (1995).

[118] M. Nestle, 'Allergies to transgenic foods – Question of policy', *N. Engl. J. Med.* 334(11), 726–728 (1996).

[119] P. Holden, 'Safety of genetically engineered foods is still dubious', *Br. Med. J.* 318, 332 (1999).

This critical view was taken up by various scientists. According to David Schubert:

> [...] the FDA has no mandatory safety approval regulation for GM foods and no specific testing requirements [...] the cited testing protocols are only suggestions for producers [...]. Parrot and colleagues claim that "the protein produced in the new host is subjected to extensive biochemical characterization to confirm that the protein produced is the one and only one intended". However, there is no technique that can assay all cellular proteins. The best to date is 2,528 out of the rice genome of 50,000 genes[120] (a mere 5%!).[121]

Despite repeated authoritative declarations by government bodies and scientists, and several attempts at amendment, the US system of controls and rules remains inadequate and incomplete. Almost 7 years after Domingo's alarm, very little has been done. Contrary to expectations of indomitable GM proponents and companies, often engaged in embarrassing operations of propaganda,[122] the implementation of control procedures, both rules and risk detection methods, not only failed to disperse doubts but has also raised new questions and problems, fuelling widespread scepticism.[123] The conclusions of the National Research Council in 2000[124] and 2002[125] underlined the inadequacy of the system of regulation. Particular concern was expressed about health risks involving transgenic animals and consumers of their meat and milk. "Laboratory experiments and other measures to simulate experimental conditions and reproduce the behaviour of these organisms will never be sufficient to predict their fate" under the current system of control that "makes it difficult, if not impossible, for the FDA to answer these questions".[126] Even stronger criticism has been directed at the fragmentation of infrastructures of control and at the dispersion of responsibility between bodies (USDA, EPA, FDA) that rarely collaborate. This situation creates confusion and inevitably gives excessive discretion to companies, on whose good faith the whole authorization procedure ultimately rests. To cite a significant example, let us recall that the biotech industry has two

---

[120]A. Kuller, M. P. Washburn, B. M. Lange, N. L. Andon, C. Deciu, P. A. Haynes, L. Hays, D. Schieltz, R. Ulaszek, J. Wei, D. Wolters & J. R. Yates, 'Proteomic survey of metabolic pathways in rice', *Proc. Natl Acad. Sci. U. S. A.* 99(18), 11969–11974 (2002).

[121]D. Schubert, *op. cit.*

[122]T. Malarkey, 'Human health concerns with GM crops', *Mutat. Res.* 544, 217–221 (2003).

[123]J. L. Fox, 'NAS report: strengthen agbio regs and relations', *Nat. Biotechnol.* 18, 486 (2000).

[124]Committee on Genetically Modified Pest-Protected Plants, Board on Agriculture and Natural Resources, National Research Council of USA. *Genetically Modified Pest-Protected Plants: Science and Regulation*, National Academy Press, Washington, DC (2000).

[125]National Research Council, *Animal Biotechnology: Science-Based Concerns*, National Academy Press, Washington, DC (2002).

[126]'US Animal Biotech Regulations "may not be adequate" ', *Nat. Biotechnol.* 20, 959 (2002).

possible courses ahead of it when initiating a field test with GM seeds: it can apply for formal authorization, in which case it has to undergo a long bureaucratic process, or it can simply notify the Animal and Plant Health Inspection Service (APHIS) of the USDA that the transgenic product meets the safety guidelines. APHIS can ask for details or rule suspension within 30 days, otherwise the request goes on its way. Obviously, in most cases (about 1,600 per year) the option of notification is chosen. If this is useful in some cases, in others, haste plays nasty tricks. In 1997 APHIS authorized, via notification, the sowing of a variety of wheat engineered to express a glycoprotein called avidin that is toxic for at least 26 species of insects, clearly violating its own guidelines.[127] Accused of superficial management of the environmental impact of GMOs, APHIS was asked to pass this responsibility over to EPA.[128]

An important report of the National Academy of Science tried to take up these concerns, suggesting new solutions. Commissioned jointly by the FDA, EPA and USDA, the report Safety of GM Foods: Approaches to Assessing Unintended Effects[129] underlined that though health consequences of consumption of GMOs had not been found, the technique used to produce these foods is new, and concern about its safety remains. The companies had conducted their own studies up to then, unrequested but undertaken spontaneously case by case, focusing exclusively on changes in composition rather than on possible impact on health. The report suggested that controls be broadened to include "unexpected effects" and extended not only to GM foods but also to foods modified by traditional practices. It however did not make anything mandatory and is likely to remain a dead letter: it does not specify the type of research to be done, nor how to adapt epidemiological surveillance, pre- and post-market, so as to detect consequences for the health of populations consuming GM foods. The task of putting these recommendations into practice is left vague, referring indistinctly to the commitment of companies and institutions. The report has been judged by informed commentators to be a Pontius Pilate operation, the indications of which are purposely enigmatic like the oracle of Delphi. Like a blanket too small for the bed, the various parties tried to pull it to their side. Sustainers of agro-biotech found implicit acceptance of the importance of GMOs, whereas opponents of transgenic foods pointed out that the text confirmed the intrinsic dangers and the need for a rigorous system of pre- and post-market controls. The report actually has several innovative points, one of which was not appreciated by the biotech corporations: in delegating the task of deciding, case by

---

[127] E. Stokstad, 'NAS asks for more scrutiny of GM crops', Science 295, 1619–1620 (2002).
[128] J. L. Fox, 'National academy panel urges USDA to toughen reviews of transgenic plants', Nat. Biotechnol. 20, 323–324 (2002).
[129] Committee on Identifying and Assessing Unintended Effects of Genetically Engineered Foods on Human Health, National Research Council, Safety of Genetically Modified Foods: Approaches to Assessing Unintended Effects, National Academy Press, Washington, DC (2004), available at: http://books.nap.edu/catalog.php?record_id=10977.

case, how and whether to broaden certifications presented by companies to the FDA and EPA, it gives these agencies the task of promoting traceability of new foods by clear and unambiguous labelling, without which any post-market surveillance programme would be impossible. According to the industry this was asking too much. The report did not explain how it was to be implemented nor who was to pay, and in their opinion this seemed unnecessary in a context where there were no apparent risks.[130]

The Biotechnology Industry Organization merely falls in line with the position of the US government, which since 1986 has strongly defended the principle that ascertainment of the impact of biotech products on health and environment does not call for special laws or specific rules. As shown by an editorial in *Nature Biotechnology* commenting on another report dated 2004 by the PEW Initiative on Food and Biotechnology, something has changed since then:

> Clearly, engineered fish or insects represent a threat to their wild relatives, requiring a close assessment of their environmental impact. There would also be no product recalls for transgenic fish, insects or bacteria released into the environment. The genies would be out of the bottle, with the potential to damage human health or the environment [...]. Not only, there are some rather awkward and potentially obnoxious genes out of there, and the regulatory stoppers in the bottles are an increasingly bad fit. The report also highlights inadequacies in the existing legislation when applied to new biotech products.[131]

Proposing better coordination between bodies or implementing risk control procedures are unlikely to solve these questions and keep up with progress in analytical technologies. A change of paradigm is needed and the concept of risk has to be rethought, though the biotech industry would probably not have accepted this ... until today.

Controversy about biological interpretations is a common path of advancement of science. Public acceptance of biotechnologies would, however, have been promoted if the debate had been settled by longer, more detailed and transparent toxicological tests on GMOs, and this should have happened 20 years ago when the most widely grown GMOs were still experimental. The most detailed regulatory tests of GMOs are 3-month feeding trials on laboratory rats, which are then assessed biochemically.[132] The tests are not compulsory, and are not conducted

[130]'NAS issues mixed message on unintended effects of GM foods', *Nat. Biotechnol.* 9, 1062 (2004). The comment was written by Lisa Dry of BIO (*Biotechnology Industry Organization*).
[131]'Playing catch-up', Editorial, *Nat. Biotechnol.* 22(6), 637 (2004).
[132]G.-E. Séralini, *Ces OGM qui Changent le Monde*, Flammarion, Paris (2004).

independently. The test data and results are kept secret by the companies. This situation can clearly no longer be accepted. Séralini and colleagues therefore "call for the promotion of transparent, independent and reproducible health studies for new commercial products, the dissemination of which implies consequences on a large scale". They stress the need for "lifetime studies for laboratory animals consuming GMOs", not the 2-year tests on rats common today for certain pesticides and drugs. "Such tests could be associated with trans-generational, reproductive or endocrine research studies. And moreover, shortcomings in experimental designs may raise major questions on other chemical authorizations."[133]

The increasing number of episodes in which the inadequacy of GMO control systems and monitoring has been demonstrated, all of which put both the biotech companies and the regulatory bodies in a bad light have fed understandable public distrust and legitimate suspicion. The tryptophan scandal,[134] contamination of StarLink maize,[135] mixing of Bt11 and Bt10 maize,[136] repeated gaffes by Monsanto about the presence of unexpected gene constructs and the toxicity of MON863[137] have not improved the public image of biotech companies and have focused attention on the need to rigorously check and control risks associated with the introduction of modified foods.[138] In the USA, this process has joined with a general critical change of mind about relationships involving the chemical and pharmaceutical industry, which has been on the defendant's bench for several years. The recent Merck Vioxx® scandal[139] demonstrated the limits of the control system: Med-Watch, which is based on voluntary reporting by doctors of secondary effects of drugs, does not intercept more than 10% of side-effects; independent research by universities are few and expensive; finally, the FDA is in a weak position compared to the industry, being able to request but not impose health risk and product safety

[133] J. S. de Vendômois, D. Cellier, C. Vélot, E. Clair, R. Mesnage & G.-E. Séralini, 'Debate on GMO health risks after statistical findings in regulatory tests', *Int. J. Biol. Sci.* 6(6), 590–598 (2010).

[134] A. N. Mayeno & G. J. Gleich, 'Eosinophilia-myalgia syndrome and tryptophan production: a cautionary tale', *Trends Biotechnol.* 12(9), 346–352 (1994).

[135] L. Bucchini & L. R. Goldman, 'StarLink corn: a risk analysis', *Environ. Health Perspect.* 110, 5–13 (2002).

[136] 'US launches probe into sales of unapproved transgenic corn', *Nature* 434, 123 (2005).

[137] Mais transgenico: la Procura di Torino chiede lo studio della Monsanto, *La Stampa*, 25 May 2005.

[138] The list of "unfortunate" and "unexpected" episodes involving "accidental" contamination of seeds and food by GMOs is long but instructive. The frequency with which they have occurred, the extent of the damage and the intolerable delay in identifying them cast doubt on their alleged accidental nature. A detailed list can be seen at http://www.grain.org/research/contamination.cfm?cases.

[139] E. Marris, 'Suppressed study raises spectre of flawed drug regulation in US', *Nature* 432, 537 (2004).

studies. Out of 1,200 requests by the FDA for studies into the secondary effects of drugs already on the market, 70% have not even begun:[140]

> The US drug safety system is outdated, weak, disorganized and seriously underfunded, according to no less an authority than the Institute of Medicine [...]. The Food and Drug Administration (FDA) has neither the money nor the muscle to police the safety of drugs already on the market [...] Vioxx provided an example of what can happen when drugs originally tested in limited numbers of people in clinical trials are put to work in the real world.[141]

More controls, further investments and an effective system of labelling to inform consumers that a drug has only recently been released without knowing all possible side-effects are needed. Faced with these facts and irrespective of good intentions, is it reasonable to trust FDA reassurance about the safety of GM foods subject to many fewer controls than new drugs?

Patrick Holden, director of the Soil Association (UK), an association of producers of fertilizer and feedstock, clearly does not think so. He has begun his own investigation of the safety of transgenic foods.[142] The results convinced the association that the safety of GM foods is still doubtful and it advised members not to use GMOs in products carrying the mark of the association. At about the same time, a measure inspired by the same common sense induced the EU to examine more closely the question of GMOs presented as "natural" or "substantially equivalent". As mentioned, EU law does not oblige labelling for foods containing less than 1% of transgenic ingredients. Though better than nothing, this rule can easily be side-stepped and is in any case ineffective. Even a tiny percentage (under 1%) can perfectly well be toxic or allergenic. The rule can also easily be ignored, since the 1% limit applies to each ingredient. Nothing prevents a food from containing 0.9% of transgenic soy, 0.9% of transgenic maize and 0.9% of transgenic lecithin. In theory, consumers could find themselves with foods that are 90% genetically manipulated and 10% organic without breaking the law.

Apart from these limits, the EU, which has always had a different attitude to GMOs from the USA,[143] saw fit to begin an international programme to check the reliability of reference laboratories that conduct tests on food for different national and international organizations. The occasion was ideal for tracing a first balance and it was discovered that more than 60% of the food offered to consumers

---

[140]'Drug safety on trial', *Nature* 434, 545 (2005).
[141]'Reforms on drug safety', *Nature* 443, 372 (2006).
[142]'Safety of genetically engineered foods is still dubious', *BMJ* 318, 332–333 (1999).
[143]G. Gaskell, 'Worlds apart? The reception of genetically modified food in Europe and the U.S.', *Science* 285, 384–386 (1999).

today contains GM ingredients (mainly transgenic soy and maize derivatives). It was embarrassing to admit that only 9/22 laboratories of excellence had quality standards sufficient for the tasks assigned them.[144] Thus, not only are we submerged in GMOs, but the quality standards of available analytical technologies are in most cases inadequate and incomplete. The USDA finally realized this and has just begun a re-examination of the laws and protocols regulating GMOs in order to consider "the whole spectrum of farm and environmental risks, including health risks, raised by GMOs".[145]

This late "second thought" negates the optimists and confirms the basis of many, often unheeded, invitations for precaution. Margaret Mellon's repeated warnings and firm stance come to mind. To all appearances a genteel middle-aged lady who passes her time between home and teaching children, Mellon is probably mild and forgiving with her family, but not in her role as director of the agricultural and biotech programme of the Union of Concerned Scientists of Washington University. As she has often stated, and as recently stressed in the columns of *The Scientific American*, Mellon's opinion is that:

> Nobody is saying, "Look, we've got this large body of peer-reviewed experimental data comparing GM with non-GM foods on a number of criteria that demonstrate the food is safe." When we have generated such a body of evidence, *then* there will be an issue of whether what we have is enough [...]. Lots and lots of people — virtually the whole population — could be exposed to genetically engineered foods, and yet we have only a handful of studies in the peer-reviewed literature addressing their safety. The question is, Do we *assume* the technology is safe based on an argument that it's just a minor extension of traditional breeding, or do we *prove* it? The scientist in me wants to prove it's safe. Why rest on assumptions when you can go into the lab?[146]

Why indeed?

---

[144]'EU examines accuracy of tests for genetically modified foods', *Br. Med. J.* 320, 468 (2000).
[145]'Biotech crop rules get rewrite', *Nature* 449, 9 (2007).
[146]M. Mellon, 'Does the world need GM foods?', *Sci. Am.* 223, 62–65 (223).

8

Bread for man?

Have we the right to counteract, irreversibly, the evolutionary wisdom of millions of years, in order to satisfy the ambition and the curiosity of a few scientists? This world is given to us on loan. We come and we go; and after a time we leave earth and air and water to others who come after us. My generation, or perhaps the one preceding mine, has been the first to engage, under the leadership of the exact sciences, in a destructive colonial warfare against nature. The future will curse us for it.

E. Chargaff

For all we have and are, for all our children's fate, stand up and take the war.

R. Kipling

Speaking of politics [...] is there anything to eat?

Totò

The Ideas of Science

In recent years, ideas about cloning and genetic manipulation have changed radically, not only in the scientific community. For a long time, this topic evoked the spectre of eugenetics and biological catastrophe, dear to science fiction. This literature was the epitome of scientific adventure in the fifties. For the first time, science has now relinquished all pretence, leaving science fiction behind it. This is a significant event: science fiction, bound by a concept of the universality and function that science and the literature that surrounds it play in the world, is reluctant to recognize betrayal of the role assigned by ethics and history to human cognitive activity. It now seems that Science (the science incarnated in institutions and humans) has burnt its bridges with the past, freeing itself of doubts and uncertainty and settling the question of its alleged neutrality: in producing modified organs and seeds it has not only begun a project of world transformation but has also submitted to the financial potentates of the so-called New Economy. After Dolly, the cloned sheep, and after the sentence in which the US Supreme Court recognized the possibility of patenting life, there is no longer any place for neutral science.

The Presumption of the New Religion

Biotechnology rests on scientific foundations and on a reductionist approach that cannot be used to understand biological complexity. For example, the central dogma of biology states that all information resides in DNA. This dogma is the basic assumption of all transgenic manipulations, namely insertion of foreign genes in a host chromosome. Discovery of the prion of bovine encephalopathy belied this assumption: other organic structures, like the prion protein, can also carry biological information. Prions have characteristics so peculiar as to cast doubt on the central dogma of biology, that DNA and RNA are the sole sources and carriers of biological information.[1] We now know that things are much more complex and that although DNA is important, it is only one of the protagonists of that complex biological system, the cell, that ensures decoding, processing and transmission of information.[2,3] The trust place in these dogmas makes these new branches of science veritable New Religions: criticism and debate degenerate on both sides because the parties do not recognize each other's arguments but label them as heresy or obscurantism against the linear and inevitable nature of progress. As happens when one believes in certain "truths" without testing them, this "new science" ends up producing monsters and disasters.

Signs of this kind have been emerging with increasing frequency in recent decades, especially in relation to technological processes that have released dangerous products into the environment. Biotechnology, including so-called gene therapies,[4] is no exception.

Evidence produced by scientific research shows that precaution and prudence are necessary in this sector, over and above pre-constituted or ideological positions. As usual, this evidence is not univocal and can often be used to sustain or deny certain positions. Moreover, every scientific truth has a slot in history and is only

---

[1] A. Aguzzi, 'Unraveling prion strains with cell biology and organic chemistry', *Proc. Natl Acad. Sci. U. S. A.* 105, 11–12 (2008).

[2] I. R. Henderson & S. E. Jacobsen, 'Epigenetic inheritance in plants', *Nature* 447(7143), 418–424 (2007).

[3] L. Van Speybroeck, G. Van de Vijver & D. De Waele (Eds), 'From epigenesis to epigenetics: the genome in context', *Ann. N. Y. Acad. Sci.* 981, 1–237 (2002).

[4] Contrary to what we read in the press, the prospects for gene therapies, especially in oncology, have stalled since the emergence of severe side-effects related to the technology and the type of vectors used to obtain gene transfer. Unexplained deaths and development of tumours in patients undergoing experimental treatment led to sudden interruption of experimental programmes previously authorized by the FDA (see F.D. Bushman, 'Retroviral integration and human gene therapy', *J. Clin. Invest.* 117(8), 2083–2086 (2007); Q. Schiermeier, 'Regulators split on gene therapy as patient shows signs of cancer', *Nature* 419, 545–546 (2002); E. Check, 'A tragic setback', *Nature* 420, 116–118 (2002); Z. Li, J. Dullmann, B. Schiedlmeier, B. Fehse & C. Baum, 'Murine leukemia induced by retroviral gene marking', *Science*, 296, 497–498 (2002)).

valid within certain well-defined limits, destined to be confirmed or denied with the advance of science. The current status of biotech does not permit anyone to exclude that its applications could have negative effects on ecosystems and human health: this is indeed a certainty, confirmed indirectly by the efforts of biotech companies to obviate the many problems encountered to date. For an idea of the confusion reigning in the sector, it is enough to realize that the exact number of genes is still unknown. The first estimates suggested about 100,000 genes in humans, amended to 66,000 and lately just over 30,000.[5] These figures continue to be revised. The fluctuations in this basic figure give an idea of the margin of approximation of any operation in this field.

Vain attempts have been made to defend GMOs at all costs, denigrating the results of other researchers and making the debate on biotechnologies out to be a dialectic between the "Seattle people" and a hypothetical monolithic "scientific community", which in actual fact is anything but monolithic in its convictions and certitudes. Discussion of the validity and safety of transgenic food is occurring largely *within* the world of science, and as on other topics, this world is revealing differences and dichotomies that are difficult to reconcile. This diversity reflects not so much divergence of evaluation and methodological approach, but opposite ideas on humans and nature. Whether we like it or not, science is no longer neutral or objective, and few people believe it to be so. C. Sonnenschein made this point very effectively when he wrote: "Philosophy is central to all scientific endeavours, including experimental and systems biology. Although many biologists ignore it, their research is guided by unstated ontological and epistemological stances. The inescapable fact is that, whether biologists like it or not, there are no theory-free data."[6] The biotech option, afflicted as it is by a reductionist and neopositivist approach, as well as heavy academic and industrial bias, is only one of many possible hypotheses for tackling the dramatic problems humanity is facing. There is no objective scientific criterion, intrinsic to science, for deciding whether or not it is the right or most economic or appropriate option.

> To understand what is right or wrong we should not knock on the door of science [...]. Science, too, needs rules and cannot develop in a wild manner or even in open defiance of current morals. A great revolution like biotechnology not only puts material aspects of our lives under the microscope, but also our culture, world-views and values. This is why it is urgent to lay the foundations for a new contract between science and society [...].

---

[5] J. M. Claverie, 'What if there are only 30,000 human genes?', *Science* 291, 1304–1351 (2001).

[6] K. Saetzler, C. Sonnenschein & A. M. Soto, 'Systems biology beyond networks: generating order from disorder through self-organization', Sem. Cancer Biol. 21(3), 165–174 (2011).

> The new contract should protect society against risks associated
> with the use of recombinant DNA.[7]

This is not unreasonable. Moreover, proponents of GM food do not reply fully to the criticism levelled against them, often avoiding any real dialogue on the major points of the controversy. It is difficult to meet and talk; there are few occasions and places where peaceful discussion could take place. Though many sustainers of biotechnologies are not insensitive to the criticism and questions raised, others (generally those occupying the media) use tones and assume positions which are either propaganda, ridiculous, or insulting. *Scientific American*, which certainly does not have an anti-GMO editorial line, has condemned such attitudes because they do not promote clarity or help to solve problems and dispel doubts.[8]

The Objections: World Hunger

GMOs are presented as the best solution to world hunger. This statement, that it would probably take another book and economic analysis to disprove, seems to deliberately ignore two elementary considerations. The first is that it forgets that the transgenic seed market is wholly in the hands of a few multinationals that are mainly interested in their own profits, unless they have become philanthropic corporations as suggested by Anna Meldolesi.[9] As in the last 6 years, capital concentration in that sector will probably continue to increase, enabling a very narrow industrial group

---

[7]R. Dulbecco, *Ingegneri della vita*, Sperling Kupfer, Milan (1988), pp. 11, 154–155, our translation.

[8]On the occasion of the recent publication of a book in favour of GMOs, M. Capocci asked if it was "legitimate to reply to *propaganda with other propaganda?* ... This is perhaps the main question that the book raises [...] facile enthusiasm and minimization of dangers expose the writer to criticism [...] The triumphalism seems inappropriate, *especially if it is calculated to cover up problems and uncertainties* [...] GMOs may not be dangerous for human health and ecology but [...] their applications are in the hands of well-known economic interests that do not work in the directions hoped by the author [...] namely alleviating world hunger and decreasing the use of pesticides (the author himself writes that 85% of GM soy is engineered to resist pesticides so that greater quantities of chemicals can be used) [...] and the problem of the unpredictable long-term consequences for the environment is unsolved" (*Le Scienze* 439, 116 (2005); our italics, our translation).

[9]A. Meldolesi, *Organismi geneticamente modificati: storia di un dibattito truccato*, Einaudi, Turin (2001). Meldolesi is so convinced of the good work of drug multinationals that she defended them at a recent meeting where the government of South Africa opposed certain giants of the world pharmaceutical industry. The spiny question was the right of the African nation to buy medicines no longer protected by patents to cure AIDS (these drugs cost one tenth the price of drugs produced by the multinationals). The fact that the government won the argument was criticized by the journalist as negative. In her words, 39 drug companies gave up the battle where defence of patents was a question of survival (A. Meldolesi, 'L'industria del farmaco fa harakiri', *Le Scienze* 395, 92–93 (2001), our translation).

to monopolize the vast market covering everything from agrochemicals to catering. We do not think that this could bring benefits to third-world countries, whose poverty springs more from unequal distribution of resources than from inability to increase productive potential.[10,11] The second is that world hunger is a tragedy of our times and has many faces and many causes. In the present climate of exasperated pragmatism, we should beware of *clichés* that acquire the status of truths. An important essay by Giovanni Monastra, director of the Italian Institute for Research into Food and Nutrition, underlines the need to approach the question in a more rational way:

> The levels of rural poverty in the so-called "developing countries" average between 50% and 70%; these people live in marginal environments [...] almost untouched by the Green Revolution, which was a real chance to improve the production only for the agricultural system of the Western farmers [...]. Furthermore, we cannot forget that most of the food production is used for feeding cattle [...] this situation is clearly abnormal and unbalanced [...] it is clear, thus, that the main problem in these areas is not the food *per se* but the need for a general improvement of the economic system [...]. Lands should be reconverted in order to mainly produce food for their inhabitants and reduce areas devoted to producing fodders for exportation. In this situation no benefit can be derived from the introduction of GM cultivations. [...] [moreover] the Green Revolution, with the introduction of new wheat and rice varieties tried to increase the production through the development of monoculture systems, with a further reduction of the diversification of food production.[12]

Though ignored by the media which deals freely in *clichés*, scholars are well aware that there is enough food to feed the whole world,[13] but it is doubtful that doubling of food resources would solve world hunger, since the new markets conceived to absorb the surplus are orientated in other directions (energy from biomass, industrial oils, medical and veterinarian applications)[14,15] that have nothing to do with food. It is also paradoxical that it is thought to solve the imbalances caused by the spread of monocultures with a new type of monoculture, that based on GM crops, especially when this facile recipe emanates from the very government

---

[10]Christian Aid, *Selling Suicide Farming, False Promises and Genetic Engineering in Developing Countries*, Report, London (1999).

[11]W. H. Verheye, 'Local farmers would be able to feed Africa if they were given the chance', *Nature* 404, 431 (2000).

[12]G. Monastra & L. Rossi, 'Transgenic foods as a tool for malnutrition elimination and their impact on agricultural systems', *Biol. Forum* 96, 363–384 (2003).

[13]S. Pouteau, 'Beyond substantial equivalence: ethical equivalence', *J. Agric. Environ. Ethics* 13, 273–291 (2000).

[14]B. Zechendorf, 'Sustainable development: how can biotechnology contribute?', *Trends Biotechnol.* 17, 219–225 (1999).

[15]J. M. Dunwell, 'Transgenic crops: the next generation or an example of 2020 vision', *Ann. Botany* 84, 269–277 (1999).

and private organizations responsible for the present dilemma in agriculture and food production, completely uninterested in assessing the potential utility of alternative development models and solutions,[16,17] including techniques of metabolic modulation not based on transgenesis.[18] These doubts and scepticism are also shared by many experts in national food institutes, such as Jean Pierre Berlan,[19] and by economists who [apart from assessments of GMO impact on health and ecology] generally agree on the insignificant productive and economic utility of GMOs.[20,21,22]

Is There Any Bargain in This?

Indeed, the claim that GM crops give higher yields is often uncritically repeated in the media. But this claim is far from accurate. At best, GM crops have performed no better than their non-GM counterparts, with Roundup soy giving consistently lower yields. A review of over 8,200 university-based soybean varietal trials found a yield drag of between 6% and 10% for GM RR soy compared with non-GM soy.[23] Controlled comparative field trials of GM and non-GM soy suggest that half the drop in yield is due to the disruptive effect of the GM transformation process.[24] Data from Argentina show that GM RR soy yields are the same as, or lower than, non-GM soybean yields.[25] Claims of higher yields from Monsanto's new generation of RR soybeans (RR 2 Yield) have not been borne out. A study carried out in five US

[16] D. Gibbs, 'Globalization, the bioscience industry and local environmental responses', *Global Environ. Change* 10, 425–257 (2000).

[17] L. Levidow, Simulating mother nature, industrializing agriculture, in: *Future Natural: Nature, Science, Culture*, G. Robertson, M. Marsh, L.Tickner, J. Bird, B. Curtis &T. Putnam (Eds), Routledge, London (1996), pp. 55–71.

[18] I. Levin, A. Lalazar, M. Bar & A. A. Schaffer, 'Non GMO fruit factory strategies for modulating metabolic pathways in the tomato fruit', *Ind. Crop Prod.* 20, 29–36 (2004).

[19] J.-P. Berlan (Eds), *La Guerra al Vivente*, Bollati Boringhieri,Turin (2001).

[20] M. Prestamburgo, Agricoltura e OGM: quale convenienza per il sistema agroalimentare italiano?, in: *Organismi Geneticamente Modificati: Il Tempo delle Scelte*, M. Bizzarri, A. Laganà & S. Vieri (Eds), Commedia, Rome (2003), pp. 71–83.

[21] C. Malagoli, 'Moderne biotecnologie e politica agraria comunitaria: un connubio difficile', *Pol Agraria* 4, 9–19 (1999).

[22] S. Vieri, Coltivazioni transgeniche: alcune valutazioni nella prospettiva della loro introduzione nel sistema produttivo agricolo italiano, in: *Agro-biotecnologie nel contesto italiano*, a cura di G. Monastra, INRAN, Rome (2006), pp. 469–489.

[23] C. Benbrook, Evidence of the magnitude and consequences of the Roundup Ready soybean yield drag from university-based varietal trials in 1998. Ag BioTech InfoNetTechnical Paper No. 1, July 13, available at: http://www.mindfully.org/GE/RRS-Yield-Drag.htm.

[24] R. W. Elmore, F. W. Roeth, L. A. Nelson, C. A. Shapiro, R. N. Klein, S. Z. Knezevic & A. Martin (2001), 'Glyphosate-resistant soyabean cultivar yields compared with sister lines', *Agron. J.* 93, 408–412 (2001).

[25] M. Qaim & G.Traxler, 'Roundup Ready soybeans in Argentina: farm level and aggregate welfare effects', *Agric. Econ.* 32, 73–86 (2005).

states involving 20 farm managers who planted RR 2 soybeans in 2009 concluded that the new varieties "didn't meet their [yield] expectations".[26]

In June 2010 the state of West Virginia launched an investigation of Monsanto for false advertising claims that RR 2 soybeans gave higher yields.[27]

A possible explanation for the lower yields of GM RR soy is that the transgenic modification alters the plant's physiology so that it takes up nutrients less effectively. In fact, one study found that GM RR soy takes up the important plant nutrient manganese less efficiently than non-GM soy.[28] Another possibility is that the new added biological function that enables the GM soy to resist glyphosate involves additional energy consumption by the plant. As a result, less energy could be left over for grain formation and maturity. The genetic engineering process permitted a new function, but did not make additional energy available.

A US Department of Agriculture report confirms the poor yield performance of GM crops: "GM crops available for commercial use do not increase the yield potential of a variety. In fact, yield may even decrease [...]. Perhaps the biggest issue raised by these results is how to explain the rapid adoption of GM crops when farm financial impacts appear to be mixed or even negative."[29] The failure of GM to increase yield potential is emphasized by the United Nations IAASTD report of 2008 on the future of farming.[30] This report, written by 400 international scientists and backed by 58 governments, says that yields of GM crops are "highly variable" and in some cases "yields declined". The report continues: "Assessment of the technology lags behind its development, information is anecdotal and contradictory, and uncertainty about possible benefits and damage is unavoidable." The definitive study to date on GM crops and yield is "failure to yield: evaluating the performance of genetically engineered crops",[31] by former US EPA scientist, Doug Gurian-Sherman. It uses data from published, peer-reviewed studies with

---

[26]J. Kaskey, 'Monsanto facing "distrust" as It seeks to stop DuPont', *Bloomberg* November 11 (2009).

[27]C. Gillam, 'Virginia probing Monsanto soybean seed pricing. West Virginia investigating Monsanto for consumer fraud', *Reuters*, June 25, available at: http://www.reuters.com/article/idUSN2515475920100625.

[28]B. Gordon, Manganese nutrition of glyphosate resistant and conventional soybeans. Better Crops 91, April (2006), available at: http://www.ipni.net/ppiweb/bcrops.nsf/$webindex/70ABDB50A75463F085257394001B157F/$file/07-4p12.pdf.

[29]US Department of Agriculture, The adoption of bioengineered crops (2002), available at: http://www.ers.usda.gov/publications/aer810/aer810.pdf.

[30]N. Beintema *et al.*, International Assessment of Agricultural Knowledge, Science and Technology for Development: Global Summary for Decision Makers (IAASTD) (2008), available at: http://www.agassessment.org/index.cfm?Page=IAASTD%20Reports&ItemID=2713.

[31]D. Gurian-Sherman, Failure to yield: evaluating the performance of genetically engineered crops. Union of Concerned Scientists (2009), available at: http://www.ucsusa.org/assets/documents/food_and_agriculture/failure-to-yield.pdf.

well-designed experimental controls. The study distinguishes between intrinsic yield (also called potential yield), defined as the highest yield which can be achieved under ideal conditions, and operational yield, the final yield achieved under normal field conditions when crop losses due to pests, drought or other environmental stresses are factored in. The study concludes that GM herbicide-resistant soybeans have not increased yields. GM crops "have made no inroads so far into raising the intrinsic or potential yield of any crop. By contrast, traditional breeding has been spectacularly successful in this regard; it can be solely credited with the intrinsic yield increases in the United States and other parts of the world that characterized the agriculture of the twentieth century". Thus, "If we are going to make headway in combatting hunger due to overpopulation and climate change, we will need to increase crop yields. Traditional breeding outperforms genetic engineering hands down".

Other aspects to consider are the cost of access to intellectual property, that threatens to exclude small farmers in developing countries,[32] and the fact that denying the possibility of risks can have very high costs, for example a million dollars for loss of market in the case of canola, and a billion dollars in penalties paid for the StarLink affair.[33] Can genetic engineering therefore provide a *technological* solution to a problem of exclusively political and economic origin? Let us hear Margaret Mellon, an impartial scientist whose authority is not doubted by anyone, though she is rarely cited by sustainers of GMOs:

> That is an important question to ask because so many people –
> more than 800 million – are undernourished or hungry. But is
> genetic engineering the best or only solution? We have suffi-
> cient food now, but it doesn't get to those who need it. Most
> hungry people simply can't afford to buy what's already out
> there even when commodity prices are at all-time lows [...]. The
> real tragedy is that the debate about biotechnology is diverting
> attention from solving the problem of world hunger. I'd like to
> see people seriously asking the question, "What can we do to
> help the world's hungry feed themselves?" and then make a
> list of answers. Better technology, including genetic engineer-
> ing, would be somewhere on the list, but it would not be at the
> top. Trade policy, infrastructure and land reform are much more
> important, yet they are barely mentioned.[34]

Actually, not even producers of GMOs still believe that GMOs will solve world hunger and are working in other directions. The modified plants on the global market are still basically only four (maize, soy, canola and cotton) and consumers'

---

[32]N. Lalitha, 'Diffusion of agricultural biotechnology and intellectual property rights: emerging issues in India', *Ecol. Econ.* 49, 187–198 (2004).
[33]S. Smyth, G. G. Khachatourians & O. W. Phillips, 'Liabilities and economics of transgenic crops', *Nat. Biotechnol.* 20, 537–541 (2002).
[34]M. Mellon, 'Does the world needs GM foods?', *Sci. Am.* December, 38–39 (2006).

perception of the benefits of GM foods is about zero;[35] in response, research has been directed to second- and third-generation GMOs to meet potential demands of affluent markets, mainly nutraceutics and plant pharmaceuticals.[36] None of these new products are yet on the market, little data has been published, and assessments of their future possible uses remain aleatory and a matter of opinion. On the other hand, introduction of biotechnologies is likely to upset the labour market in agricultural production, redistribution of resources in industrial countries and between the latter and developing countries: nobody yet has a convincing answer.[37] Contrary to elementary common sense, the media are ventilating the idea of vaccines against cancer, hepatitis and AIDS, obtained from GM plants. It seems a promising line of research and ensures funding and fame to dedicated researchers, however, the results are still few and disappointing, and prudence is therefore necessary. To avoid a distorted picture and unfounded hope among patients:

> it would perhaps be better to wait until a vaccine against HIV [obtained by genetic modification of a plant] is actually discovered before writing that plants producing vaccines against infectious diseases and tumours (cholera, hepatitis B, AIDS, melanoma) are in an advanced stage of experimentation.[38]

Second Objection: Regulation and Control

Meldolesi states confidently that:

> Since it is commonly said that GM products will be marketed without passing severe tests, I would like to point out that producers have to present the results of a long series of tests in order to obtain approval for marketing. First of all they have to compare the transgenic variety with conventional varieties of the same species to verify that the genetic engineering did not change the metabolism of the plant or the nutritional value of the product [...] toxicity is examined by nutritional studies on guinea pigs that are fed with GM foods [...] To test allergenicity it is also common practice to compare the structure of the protein expressed after genetic modification with that of about 500 known allergens.[39]

We have already seen that these measures are not enough and that technically, pre- and post-market risk assessment and monitoring of unexpected effects is not an

---

[35] J. M. Dunwell, 'Future prospects for transgenic crops', *Phytochem. Rev.* 1, 1–12 (2002).
[36] A. Mercenier, U. Wiedermann & H. Breitneder, 'Edible genetically modified microorganisms and plants for improved health', *Curr. Opin. Biotechnol.* 12, 510–515 (2001).
[37] P. de Puytorac, 'Biotechnologies. Conséquences socio-économiques', *Année Biol.* 39, 123–204 (2000).
[38] M. Capocci, 'Gli OGM sono davvero pericolosi?', *Le Scienze* 439, 116 (2005), our translation.
[39] A. Meldolesi, *op. cit.*, p. 65, our translation.

easy problem to solve. These tests are not always conducted or are conducted using inappropriate parameters and models. Meldolesi implies that the intention of the law is actually translated into practice: in other words, we are in the best of all possible worlds. In support of her arguments, she inappropriately cites the case of Pioneer Hi-Breed that declined to release a variety of engineered soy because during pre-market testing the researchers found evidence of allergenicity,[40] implying that disaster was avoided thanks to tests conducted by Pioneer. In actual fact the studies were conducted by Prof. Nordlee's group at the University of Nebraska. They were brought to the attention of the FDA and consumers by an alarmed editorial in the *New England Journal of Medicine*, which Meldolesi does not cite. The editorial stated that the soy in question, though intended for animals, could not easily be separated from that prepared for human consumption. Because soy proteins are less allergenic than those of milk, they are used in baby formula, precooked foods, nutritional supplements and milk substitutes, but the allergenic potential of proteins obtained from GM foods is uncertain, unpredictable and untestable.[41] Moreover, the work of Nordlee's group concerned modified soy, but the same type of genetic transformation was also carried out on tobacco,[42] canola,[43] *Vicia narborensis*[44] and beans,[45] without anyone ever raising doubts about the safety of the operation. Did anyone at Pioneer take the trouble to check whether any of these products expressed allergenic properties? We found no evidence of any such studies conducted by independent scientists,[46] before or after, despite the recommendations of the journal. Examples of this fact end up having repercussions on proponents of biotech. In fact, strong criticism of the protocols used to evaluate new GMOs was published by the very journal in which Meldolesi writes, a journal

---

[40]A. Meldolesi, *op. cit.*, p. 66.

[41]M. Nestle, 'Allergies to transgenic foods – question of policy', *N. Engl. J. Med.* 334(11), 726–729 (1996).

[42]S. B. Altembach, 'Enhancement of the methionine content of seed proteins by the expression of a chimeric gene encoding a methionine-rich protein in transgenic plants', *Plant. Mol. Biol.* 18, 235–240 (1992).

[43]P. Guerche, 'Expression of the 2s albumin from *Bertholletia excelsa* in *Brassica napus*', *Mol. Gen. Genet.* 221, 306–310 (1990).

[44]S. B. Altembach, Accumulation of a Brazil nut albumin in seeds of transgenic canola results in enhanced levels of seed protein methionine, *Plant. Mol. Biol.* 13, 513–517 (1989).

[45]F. J. L. Aragao, 'Particle bombardment-mediated transient expression of a Brazil nut methionine-rich albumin in bean (*Phaseolus vulgaris*)', *Plant. Mol. Biol.* 20, 357–361 (1992).

[46]Independence of judgement is obviously a necessary but not sufficient condition for studies to be credible. Little can also be hoped from research conducted in centres under the direct control of biotech companies. Indeed, the take-over of many small and medium biotech companies by large multinational groups has been accompanied by a contraction in investments and in research staff, followed by a drop in innovation due also to the loss of competition that occurs with any concentration of industry (see D. E. Schimmelpfennig, C. E. Pray & M. F. Brennan, 'The impact of industry concentration on innovation: a study of US biotech market leaders', *Agric. Econ.* 30, 157–167 (2004)).

by no means favourable to environmentalists. In *Scientific American* news, the conclusions reached at the University of Leiden were reported:

> Authorization to deliberately release GMOs into the environment for commercial purposes or research should be subject to demonstration that they are innocuous for humans and ecosystems. According to an analysis conducted by a research group of Leiden University, The Netherlands, the currently used procedure is inadequate from this point of view. The Dutch study was financed by the *Nederlandse Organisatie voor Wetenschappelijk Onderzoek*, the principal research organization of that country. Its criticism of the current system was that the definition of the dangers of GMOs is practically determined by the applicant, who is inherently interested in obtaining authorization and therefore cannot be neutral in providing data that could influence approval. When the form provided by the Dutch government was examined, the terms in which the questions were formulated were found to be generic, leaving the decision of what information was significant for risk assessment to the applicant. For example, where the applicant is asked to explain the differences between the GMO and the original organism or whether the modifications could spread in the environment, the answer may be given in an elusive way. According to the Leiden researchers, the competent authorities should provide a series of questions designed to investigate the dangers of GMOs at all levels of analysis. It is necessary to examine the question at every step (agent, effect, context, etc.) and identify the major factors in each phase of the process. The importance of the elements to assess must be discussed and decided by experts (molecular biologists, ecologists, etc.) as well as by the parties involved (applicant and authorizing body). The importance of this aspect has been underestimated by the scientific press and also neglected in recent directives and policies of the EU (especially 90/220/CEE) and the Organisation for Economic Cooperation and Development.[47]

This position is completely in line with what emerged in the study conducted by the US National Research Council, which showed grave irregularities in the control procedures required by EPA of companies producing GMOs. The report of the committee reveals that the FDA requires companies to provide the results of nutritional and toxicological analysis only if it considers that the new plant may be unsafe,[48] leaving it to industry's discretion whether or not to do the tests. The committee found that long-term toxicity tests were not done in many cases; *these tests are required for new plant varieties obtained by traditional hybridization*, and have often revealed that new variants of foods affected the behaviour of the animals

---

[47] 'Notiziario', *Le Scienze* February 2 (2000), our translation.
[48] Committee on Genetically Modified Pest-Protected Plants, Board of Agriculture and Natural Resources, National Research Council of USA. *Genetically Modified Pest-Protected Plants: Science and Regulation*, National Academy Press, Washington, DC (2000), p. 29.

studied.[49] Secondly, some tests were found incongruous, for example the one used to assess long-term toxicity of Bt wheat, which incredibly, was tested by adding it to the diet of catfish! It is almost pathetic to point out that catfish are not a physiological model for humans; the most appropriate model would be mammals that feed on grain or forage.[50] However, the committee's severest criticism regarded the so-called *categorical exemptions*: these are the transgenic varieties exempted from the requirement of toxicity test documentation by EPA.

> The committee questions the scientific basis of the categorical exemption of plant-pesticides from sexually compatible plants and EPA's rationale. Although the committee agrees that there are few documented cases of new plant cultivars causing food safety problems (point 1), the committee does not believe that this provides a scientific basis for a categorical exemption of plant-pesticides from sexually compatible plants in light of the examples provided in this report. EPA's third point is questioned on the basis of evidence of indirect effects on nontarget organisms and data on the persistence of some naturally-occurring plant secondary compounds [...]. EPA's points 4 and 5 are questioned because transgenic methods can create a situation where a gene product is not regulated by the normal regulatory systems in the plant (for example, use of constitutive promoters). [...] that there is not sufficient data on chronic effects on humans (point 2), and that some of these compounds (for example, alkaloids) share a similar mechanism of activity as do organophosphates [...] lack of experience with transgenic pest-protected products and public concern with these products constitute practical reasons for not granting a categorical exemption to transgenic pest-protectants derived from sexually compatible species.

In summary, the committee recommended that

> Given that transfer and manipulation of genes between sexually compatible plants could potentially result in adverse effects in some cases (for example, modulation of a pathway that increases the concentration of a toxicant) and given the public controversy regarding transgenic products, EPA should reconsider its *categorical* exemption of *transgenic* pest-protectants derived from sexually compatible plants. [...] the committee questions the categorical exemption of all viral coat proteins under FIFRA due to concerns about outcrossing with weedy relatives. [...] EPA should not categorically exempt viral coat proteins from regulation under FIFRA. Rather, EPA should adopt an approach, such as the agency's alternative proposal [...], that allows the agency to consider the gene transfer risks associated with the introduction of viral coat proteins to plants.[51]

---

[49]Committee on Genetically Modified Pest-Protected Plants, *op. cit.*, p. 120 (our italics).
[50]Committee on Genetically Modified Pest-Protected Plants, *op. cit.*, p. 119.
[51]Committee on Genetically Modified Pest-Protected Plants, *op. cit.*, p. 133.

The report speaks out even more strongly on the topic of hormones produced by genetically manipulated plants. Again, EPA arbitrarily exempted producers from documenting the safety of their products:[52]

> Plant hormones often cause multiple changes in plants, including changes in secondary metabolites that might be toxic, so the scientific basis of such an exemption is questionable. [...] the categorical exemption of substances that act through non-toxic modes of action mostly considers human health effects. [...] there is a need to consider separately the impact of such substances on non-target species and the potential for the genes that code for these substances to move to feral populations or weedy relatives of the crop, where they could increase recipient plants' fitness. Categorical exemption under FIFRA might not be scientifically justifiable.

Thus, by explicit admission of the Committee of the National Research Council, minor and major risks for human health *do* exist and must be carefully assessed since there are no controls, despite what GMO producers tell us. We find the same situation for ecological risk, namely the possibility that spread of GM seeds has a negative impact on ecosystems. The committee also recommended that research:

> [...] determine the impacts of specific pest-protected crops on non-target organisms, compared with impacts of standard and alternative agricultural practices through rigorous field evaluations. Gene flow between cultivated crops and wild relatives was the second ecological impact considered by the committee. On the basis of the literature, the committee found that pollen dispersal can lead to gene flow among cultivated crops and from cultivated crops to wild relatives [...] the committee found that the transfer of either conventionally bred or transgenic resistance traits to weedy relatives potentially could exacerbate weed problems, but such problems have not been observed or adequately studied. Therefore, the committee recommends further research to assess gene flow and its potential consequences: develop a list of plants with wild or weedy relatives in the United States; identify key factors that regulate weed populations; assess rates at which pest resistance genes from the crop would be likely to spread among weed populations; and evaluate the impact of specific, novel resistance traits on the weed abundance.[53]

We recommend everyone to read the report. It is worth more than a lot of talk.

Much Rice for Little Vitamin A

The example of Golden Rice is a good illustration of the limitations of the reductionist approach of molecular biology, and of the artful media spin that new wave

---

[52]Committee on Genetically Modified Pest-Protected Plants, *op. cit.*, p. 132–133.
[53]Committee on Genetically Modified Pest-Protected Plants, *op. cit.*, p. 9 *et seq.*

biotechnology enjoys. In 2000, *Science* reported that rice had been engineered to produce beta-carotene (pro-vitamin A).[54] The announcement was accompanied by intense publicity praising the miraculous qualities of the new variety that "could save a million kids a year"[55] from blindness. The problem is severe in Africa where malnutrition and vitamin A deficiency threaten the sight of about 3 million children per year. It was immediately evident that the GM variety expressed too little beta-carotene, so in 2005, after repeated announcements,[56] Syngenta presented a new variety of rice with higher beta-carotene levels: about 30 μg/g of rice. This is an important result, but the daily requirement of vitamin A is about 700 mg, which means that about 3 kg of rice (uncooked) per day would be needed. The paradox is grotesque:[57] one fights malnutrition by modifying a component genetically but also increasing consumption of a food out of all proportion. Whoever can buy 3 kg of rice can also buy one or two carrots to meet beta-carotene requirements! It is also doubtful that Golden Rice would be accepted by those for whom it is designed: the poor could not pay the price[58] and the change in colour and flavour makes the rice less attractive. Not only: a large proportion (63%) of the beta-carotene is lost during cooking and storage,[59] while uptake by the body depends on dietary availability of zinc and other cofactors and nutrients.[60] As many scientists who are not adverse to biotechnologies have observed, "Golden rice would deliver amounts of vitamin A that are modest, and unlikely to fulfil requirements."[61] Anyone with a knowledge of biochemistry realizes that Golden Rice is good for nothing. According to Richard Lewontin,[62] Golden rice does not provide vitamin A but is enriched with a precursor of the vitamin, beta-carotene, which is converted to vitamin A in the body. But for this to happen, the eater must already have good

---

[54]X. Ye, S. Al Babili, A. Kloeti, J. Zhang, P. Lucca, P. Beyer & I. Potrykus, 'Engineering the provitamin A (betacarotene) biosynthetic pathway into (carotenoid-free) rice endosperm', *Science* 287, 303–305 (2000).

[55]This rice could save a million kids a year. *Time Magazine* July 31 (2000), vol. 156(5).

[56]Syngenta Media Release, Syngenta to donate Golden Rice to Humanitarian Board, 14 October (2004), available at: www.syngenta.com.

[57]Evidently nobody realized the enormity of this claim: 3 kg of uncooked rice is equivalent to almost 7 kg of cooked rice, to eat per day!

[58]It is estimated that the extra cost of modified rice would be the same as the cost of simply adding vitamin A to the normal grain eaten (see D. Dawe, R. Robertson & L. Unnevehr, 'Golden rice: what role could it play in alleviation of vitamin A deficiency?', *Food Policy* 27, 541–560 (2002)).

[59]F. S. Solon, R. D. W. Klemm, L. Sanchez, I. Darton-Hill, N. Craft & P. Christian, 'Efficacy of a vitamin A-fortified wheat-flour bun on the vitamin A status of Filipino schoolchildren', *Am. J. Clin. Nutr.* 72, 738–744 (2000).

[60]J. J. M. Castenmiller & C. E. West, 'Bioavailability and bioconversion of carotenoids', *Ann. Rev. Nutr.* 18, 19–38 (1998).

[61]D. Dawe, R. Robertson & L. Unnevehr, 'Golden rice: what role could it play in alleviation of vitamin A deficiency?', *Food Policy* 27, 541–560 (2002).

[62]R. Lewontin, *It Ain't Necessarily So: The Dream of the Human Genome and Other Illusions*, 2nd Revised Edition, Granta Books (16 October 2001).

nutrition and the inventors of Golden rice do not mention this in their advertising. They admit that:

> Definitive statements on the benefit of Golden Rice for the alleviation of vitamin A deficiency cannot be made. The vitamin A delivered and its impact on the body depends on several unquantified factors, including beta-carotene uptake and conversion to vitamin A, as well as the amount of rice consumed by the individual.[63]

The cultivation of Golden Rice also poses ecological problems related to spread of the transgene (rice has a high frequency of cross-pollination) and emergence of a high rate of phenotypic variability.[64] Unexpected effects surprised researchers from the beginning, when a red variety due to lycopene was expected, whereas a yellow one due to carotene was obtained. From the safety point of view it makes no difference, but it took the scientists 5 years to understand and explain it.[65]

The case of Golden Rice was used as an icon of the new biotechnologies and advertised as an example of the miracles to be expected. It has now become emblematic of how little is known of the physiology and complexity of plants. The affair also illustrates the absurdity of trying to solve medical (blindness due to vitamin deficiency) and social (child hunger and malnutrition) problems by genetic engineering when there are much more reasonable and concrete options[66,67] that are cheap and do involve meat or food supplements.[68]

---

[63] J. A. Paine, C. A. Shipton, S. Chaggar, R. M. Howells, M. J. Kennedy, A. L. Silverstone & R. Drake, Improving the nutritional value of Golden Rice through increased pro-vitamin A content', *Nat. Biotechnol.* 23, 482–487 (2005).

[64] K. Datta, N. Baisakh, N. Oliva, L. Torrizo, E. Abrigo, J. Tan, M. Rai, S. Rehana, S. Al-Babili, P. Beyer, I. Potrykus & S. K. Datta, 'Bioengineered "golden" indicates rice cultivars with beta-carotene metabolism in the endosperm with hygromycin and mannose selection system', *Plant Biotechnol. J.* 1, 81–90 (2003).

[65] P. Schaub, S. Al-Babili, R. Drake & P. Beyer, 'Why is Golden Rice golden (yellow) instead of red?', *Plant Physiol.* 138(1), 441–450 (2005).

[66] A study conducted in Bangladesh demonstrated that a daily intake of 75 g of Indian spinach, available all year round, provides enough vitamin A to meet daily requirements (cf. M. J. Haskell, K. M. Jamil, F. Hassan, J. M. Peerson, M. I. Hossain, G. J. Fuchs & K. H. Brown, 'Daily consumption of Indian spinach (*Basella alba*) or sweet potatoes has a positive effect on total-body vitamin A stores in Bangladesh men', *Am. J. Clin. Nutr.* 80, 705–714 (2004). The retinol-equivalent of a carrot is greater than that of any food of animal origin (except liver) and many other plants of tropical and subtropical zones are excellent sources of vitamin A, with concentrations up to 10 times that found in eggs or liver (R. McCance & E. Widdowson, *The Composition of Foods*, 6th Edition, Food Standards Agency, Royal Society of Chemistry, Cambridge, MA (2002)).

[67] MI (Micronutrient Initiative) & UNICEF, Vitamin & Mineral Deficiency. A Global Progress Report. Micronutrient Initiative, Ottawa, Canada (2004), available at: http://www.micronutrient.org.

[68] G. Cannon, 'Nutrition: the new world map', *'Asia Pacific J. Clin. Nutr.* 11(Suppl.), S480–S497 (2002).

Behind this technological madness, there is more than just underestimation of the problem: there is an evident lack of common sense and reasoning:

> The people using (these) techniques, and the people authorizing the funding for their introduction, seem to have lost all sense of biology and of complexity of living things.[69]

Perhaps this is what should worry us ….

Do Not Open that Door

Do we have to wait years until the consequences of insane experiments, conducted on a large scale, become evident? This is the idea with GMOs, but we forget that we are the guinea pigs. Once the new genes have been released into the ecosphere, they cannot be recalled. The risk associated with "experiments" of this kind is intrinsically unquantifiable and therefore eludes any type of cost–benefit analysis. Irreversible experiments of this kind are unacceptable, especially when the declared aim is to construct chimeras and soulless monsters. This can only be justified by our unscrupulous economic "philosophy", in which only the "market" matters. Suppressed by modern culture that is already beyond the pale, "monsters" are by definition outside the order of biological development and violate its categories. These monstrosities reveal that the overall plan behind the technical artifice is to destroy categories and the notion of species as a distinct, identifiable and essential entity of natural heritage. Without these categories, nature becomes a sort of laboratory in which life is "recreated" all the time, by the hand of man, the new demiurge. Reduced to a bank of potentially transferable genes over which patentees exercise totalitarian power, this man-made life lacks sense, beauty and equilibrium.

But even worse things are happening, as Prof. Pietro Perrino, director of the CNR Gene Bank in Bari, can tell us. Despite administrative neglect, lack of funds and behind-the-scenes manoeuvres by pressure groups to turn off the refrigerators where the seeds were kept, Perrino saved the germplasm stored in the Bari gene bank single-handedly.[70] Gene banks were established a few decades ago to collect, conserve, characterize and appreciate the genetic diversity eroded by the first

---

[69] D. F. Horrobin, 'Innovation in the pharmaceutical industry', *J. R. Soc. Med.* 93, 341–345 (2000). Regarding lack of common sense, there are many examples of cases in which science has been importuned with silly questions: experiments to produce featherless chickens, pear-shaped tomatoes and tearless onions. The case of onions is instructive. Apart from the fact that there are many ways of avoiding shedding tears while cutting up onions (just ask any housewife), one wonders if this "problem" justifies the deployment of teams of researchers and investments that could have been used more fruitfully.
[70] After a long legal battle, Prof. Perrino was nominated legal keeper of the Bari Plant Gene Bank on the basis of a decree of the court that found negligence and violation of security such as to threaten conservation of the germplasm.

"Green Revolution" that increased agricultural production through monocultures of few varieties and few agricultural species. Bioengineers are suspiciously interested in the gene pool conserved in these banks because it offers sequences with which they could produce new varieties. The danger is real:

> We have to defend the gene bank and the stored plant genetic resources from [those] who want to have full control of the germplasm in order to use it as a pretext for getting large research funds, as they are not interested in biodiversity but are fully involved in GMOs or even worse [...], the lobby, linked to multinationals, wants to destroy the germplasm and prevent farmers from using it.[71]

Episodes like this reveal the guiding light of the apocalyptic new religion: all lives boil down to their DNA; they are nothing but a soulless and spiritless toolbox that expresses functions and programmes, computers in which combinations and permutations of DNA make it possible to re-invent life. While we are at it, why not perfect life, since Man is thought to be a better architect than God?

This arrogant presumption resembles the secret desire of some alchemists.[72] It is a striking statement that contains an error and a truth. The *error* resides in the fact that true alchemists did not work with material processes, which was the domain of "puffers and blowers", but endeavoured to transcend and transform (meaning to *go beyond the forms*) the "base matter" or "lead" of their own imperfect being. The process was calcination in a crucible (athanor) consisting of initiation steps to acquire virtues, or "tools" to awaken the spiritual nature slumbering in Man. However, alchemists did not interfere with the order of things, nor did they re-design Nature, which they could only command by "learning to obey her".[73] The *truth* lies in the fact that, at least from a certain period on, under the influence of the Renaissance and then the Enlightenment, a Hermetic line emerged that neglected higher principles and led alchemy in search of the philosopher's stone, a narcissistic and hedonistic amplification of human powers, first and foremost to ensure immortality. It is here that modern science diverged from the ancient "science" of alchemy. Rightly, Simone Weil recalls that men of science and esoteric questions had a feeling in common:

> Pure virtue and wisdom [were] a necessary condition for the success of their operations, while on the other hand Lavoisier sought a recipe to combine oxygen and hydrogen to make water that would work as well in the hands of an idiot or a criminal

---

[71] P. Perrino, 'Italy's Genebank at Risk', *SiS 27* 35, 21–25 (2007).
[72] J. Rifkin, *The Biotech Century: Harnessing the Gene and Remaking the World*, Tarcher, New York (March 23, 1998).
[73] M. Maier, *Atalanta fugiens, Oppenheim 1618*. Regarding the relationship between science and alchemy; see M. Bizzarri, *Dall'Alchimia alla Chimica*, Edition Universitarie, La Sapienza, Rome (2004).

> as in his own hands [...thus it happened that] science ended up
> formulating recipes that only criminal idiots could think of using:
> technical-scientific development selects this kind of specialist.[74]

This is not new and it is not the product of that Enlightenment accused of the ills
of the modern world by a certain "traditionalism". It goes further back to certain
"deviant" lines of the Italian and especially English Renaissance, whose intricate
affairs and ambiguous overlaps with the background of occultism were described
with great elegance by Francis Yates.[75] In the early 1600s, an epigone of John Dee
and the alchemists of Prague wrote:

> We have also large and various orchards and gardens [...] And
> we make (by art) in the same orchards and gardens, trees and
> flowers to come earlier or later than their seasons; and to come
> up and bear more speedily than by their natural course they do.
> We make them also by art greater much than their nature; and
> their fruit greater and sweeter and of differing taste, smell,
> colour, and figure, from their nature. And many of them we so
> order, as they become of medicinal use. We have also means to
> make divers plants rise by mixtures of earths without seeds; and
> likewise to make divers new plants, differing from the vulgar;
> and to make one tree or plant turn into another.[76]

This resembles an advertisement of a multinational agricultural input supplier, but
the writer was actually "Father of Salomon's House", who in Francis Bacon's book
The New Atlantis, plays the role of "lord of research". The text is about a society
based on the ideals of a utopian technocracy, where the problems of humanity
have been solved by science. Probably not even Bacon believed that this was the
real intention, since he had just candidly admitted that "The end of our foundation
is the knowledge of causes, and secret motions of things; and the enlarging of the
bounds of human empire, to the effecting of all things possible".[77]

Modern technological enterprise rightly hails back to Bacon, especially where
he indicates acquiring power over nature (considered malign and imperfect) for the
needs of the few (the oligarchy of the New Atlantis) which in all this see nothing
but the possibility of more power and profits.

The epistemologist Pierre Thuillier observes:

> a science reduced to a technique or artificial manipulation of
> life, a science that deprives the world of its soul and vital force

---

[74]Encyclopédie des nuisances, *Osservazioni sull'agricoltura geneticamente modificata e
sulla degradazione delle specie*, Bollati Boringhieri, Milan (2000), p. 19 (our translation).
[75]F. Yates, *The Rosicrucian Enlightenment*, Routledge & Kegan Paul, London (1972).
[76]F. Bacon, *The New Atlantis*, 1626, available at: http://www.fullbooks.com/The-New-
Atlantis.html.
[77] *Ibid.*

and that denies or hates nature, such a science is part of the disease afflicting our scientists and technicians at this end of millennium.[78]

For Thuillier's scientists and technicians, the living world is *in power* and organisms are not distinct entities whose existences have dignity and need of the great orchestra of creation, but simply a container of resources – transferable genes – on which science and industry can claim exploitation rights. Likewise, borders between species are no longer a *sign* of identity to conserve, being the foundation of the mutual relations on which ecosystem equilibrium is based, but convenient labels that can be changed at a whim, just as children's constructions are built and dismounted. This is why bioengineering is in one sense a pervasive and all-inclusive *philosophy* of intensive exploitation of nature, which excludes all other thought forms in the name of science; in another sense it is a *practice*, legitimized by theoretical assumptions, capable of transforming reality but without the knowledge or moral right to do so.

Its techniques are methods of degeneration of life and are harbingers of disaster because they exclude conscience and responsibility. It is therefore legitimate to say that *future material damage is somehow written in the spiritual damage already present*. How else can we condemn a science that aims to modify nature and humans themselves, by reprogramming invisible genes through blind bombardment and manipulation? It is unknown where the transgene will lodge, whether and how the gene sequence will be expressed, and whether or not it will undergo fragmentation or be accompanied by multiple cell damage. It is not known whether in so doing other pieces of DNA are transferred or how they can be transmitted to other organisms or generations, and whether the characters they encode can have unintended effects, some of which could be dangerous. We proceed like Bruegel's blind, towards the abyss, completely unaware of the medium- and long-term consequences. Even the immediate results are cynically and calculatedly left to chance. Dolly, the first cloned sheep, emblem of the world that awaits us, was only obtained after 278 attempts involving carpet bombing instead of intelligent methodology. Nobody knows how a "positive" result was finally obtained, nor anything about the abortive attempts, the anomalies induced by the failed experiments, and the purposeless suffering of the animals. But why worry about a life whose identity is diluted and confused by being reduced to economic value?

This phenomenology of *absence of spirit* rests on deception, self-propagated by a refusal to face real data produced by science, treating it as insignificant or an exception when it does not go in the direction the experiment was meant to go. The case of Dr Pusztai, who lost his job because he denounced the dangers of transgenic potatoes, is emblematic for two reasons. First, it demonstrates that in

---

[78] P. Thuillier, La génétique menace-t-elle l'alimentation?, in: *Science et Vie* (1996), n. 950 (our translation).

order to deny the truth of an inconvenient result, the result is made the target of malicious and captious criticism, which has the opposite effect to that intended, antagonizing the scientific community. Second, it demonstrates the type of idea some have of "freedom of research": if it means obtaining more public money and total impunity, even witch hunts are acceptable, but if it means defending the image of a corporation on whom the result would cast doubt, then that freedom can be sacrificed. The unlucky researcher would best pack his bags and go elsewhere. Thus while in Italy someone asked Parliament for freedom of experimentation because "GMOs never killed anyone", in Scotland, Pusztai was suspended, his contract not renewed and his research group disbanded. He must have abused that freedom!

This should not surprise us. The "new science" is purposely totalitarian: it rejects scientific elements that contradict it, it purports to be the only knowledge, and it imposes its ethic while denying rights and voice to all other ethics. Its motto is simple: the only ethic is an *absence of ethics*. We are not far from the situation described by the occultist Crowley: "Do whatever you want, let this be your only law". Provided, of course, that the final result makes money. As observed by Bertrand Russell, science as such cannot provide an ethic. It can only indicate ways of achieving certain objectives and tell us what objectives cannot be achieved. According to Renato Dulbecco, it is useless to knock on the door of science: for ethics one must turn to moral philosophy or religion.[79] Precious words, too often forgotten and derided by the new apprentice sorcerers. This is why it is difficult to find common ground with these "salaried technicians" that pose as scientists, solely to denounce the "obscurantism" of their opponents, but who are actually only

> [...] degenerate descendants of men of science [...] examples of the degradation of the species they have caused. The precept crystallized in their technique is not scientific but military (they are fighting a war): they forge ahead and will concern themselves with the results later.[80]

When as students we ventured into the fascinating and phantasmagorical world of molecular biology, our heads full Jacques Monod,[81] things were not yet like this. We all still had the sensation that we were the first to venture into a sacred place, never previously violated, well aware that we had to decipher secrets that would open the doors of knowledge, oblivious of any practical application, even the most noble. That ingenuous phase was short-lived. The same presumably happened to Paul Berg, Nobel laureate, among the fathers of biotechnology, who presided the first committee of sages nominated by the US Academy of Science, with the task

---

[79]R. Dulbecco, *Ingegneri della vita*, Sperling & Kupfer, Milan (1988), p. 11.
[80]Encyclopédie des nuisances, *op. cit.*, pp. 22–23.
[81]J. Monod, *Chance and Necessity*, Vintage Books (September 12, 1972).

of assessing the dangers of the new technologies. Concern arose regarding the first modified bacterium, *Escherichia coli*, a microorganism normally found in the colon and very susceptible to becoming virulent. The fear was production of microbes resistant to antibiotics that could trigger uncontrollable epidemics. Paul Berg called an international conference at Asilomar in 1975. The meeting went down in history: it was decided that the manipulation of microorganisms had to take place in laboratories from which no bacteria could escape and that the bacteria studied had to be "weakened" so that they could only survive under special laboratory conditions. That was the first and last conference in which molecular biologists spontaneously underwent a form of self-control.[82] "Within a few years, many of those who had promoted the safety campaign changed their minds, invoking lenience and resumption of activity."[83] What happened was that many of these illustrious scholars, following the example of Paul Berg himself, founder of *Genentech*, the first biotech company in the world, simply joined the new companies, attracted by the mirage of power and money. There has never been another Asilomar. In this way two traditional barriers that at least apparently isolated science and scientists fell: the separation between the real world and the laboratory, that was broken to transform the whole earth into one big field experiment; and the separation between the academic world and big industry, which is why it is no coincidence that the same people now occupy important positions in both sectors. Thus it became possible to bend the direction of research to the exclusive advantage of the profits of big industrial groups. The myth of the neutrality of science was dead and the techno-agrochemical and agribusiness complex was born.

Other obstacles also had to be removed, firstly the theoretical constraint that links "the part with the whole". Despite the emergence of holistic disciplines, such as ecology and systems biology, biotechnology took the neopositivist paradigms of late 19th century microbiology and biochemistry to their extreme consequences. To understand phenomena, it was enough to consider them in isolation, cutting them out of the multifactorial chain of their contexts to make them seem causes and effects. This is what philosophers call the "theorem of conservation of continuity", by which events are described as more or less linearly dependent on each other: by controlling the second, one can know and programme the first deterministically. In other words, if a function is expressed by a protein coded by a gene, then by modifying the gene one can modify the protein and hence the function. This representation is a joke.[84,85] After the studies of Thom on multifactorial

[82]S. Krimsky, 'From Asilomar to industrial biotechnology: risks, reductionism and regulation', *Sci. Cult. (Lond)* 14(4), 309–323 (2005).
[83]R. Dulbecco, *op. cit.*, p. 19.
[84]R. Gallagher & T. Appenzeller, 'Beyond reductionism', *Science* 284, 79–80 (1999).
[85]B. C. Goodwin, *How the Leopard Changed its Spots*, Princeton University Press, Princeton, NJ (1994).

complexity[86], the ubiquity of non-linear dynamics[87,88] and the role of chaos in vital processes,[89] after the advances of non-equilibrium thermodynamics[90,91] and molecular biology[92,93,94] – by virtue of which we know that DNA is not enough to define a phenotype, but that the same DNA can be related to different phenotypes[95] – and after the overwhelming revelation of the prion of mad cow disease, that reminds us that the central dogma of biology is "bullshit",[96] after all this, it is astounding that someone still believes the false promises and is building "possible worlds" on them.

It is not clear how this dichotomy imposed itself. Much depends on the incongruous equation between knowledge and creation established by mechanistic philosophy. In Lenin's *Materialism and Empirio-Criticism* we read:

> If we can prove the correctness of our conception of a natural phenomenon by *creating it ourselves*, producing it from its conditions and – most importantly – *making it serve our ends*, we will finally have finished with Kant's unknowable "thing in itself" [...] The question of whether human thought can reach objective truth is not a theoretical question, *but a practical problem. It is in practice that man has to demonstrate [he possesses] truth, i.e. the reality and power, the 'this-sidedness' of his thinking.*[97]

In other words, "we demonstrate that we know what a drop of water is by reproducing one". Actually, we have no criteria for determining if the two drops of water are "really" the same and we mistake the appearance of what we reproduce with its essence. For the same reason, the much vaunted criterion of "substantial

---

[86] R. Thom, *Stabilité structurelle et morphogénèse. Essai d'une théorie générale des modèles*, Benjamin, Reading Mass, Paris (1972).

[87] Z. Yoshida, *Non Linear Science. The Challenge of Complex Systems*, Springer-Verlag, Berlin (2010).

[88] A. L. Barabasi & Z. N. Oltvai, 'Network biology: understanding the cell's functional organization', *Nat. Rev. Genet.* 5, 101–113 (2004).

[89] D. S. Coffey, 'Self-organization, complexity and chaos: the new biology for medicine', *Nat. Med.* 4, 882–883 (1998).

[90] I. Prigogine, *Introduction to Thermodynamics of Irreversible Processes*. Wiley, New York (1962).

[91] D. K. Kondepudi, 'Detection of gravity through nonequilibrium mechanisms', *ASGSB Bull.* 4, 119–124 (1991).

[92] A. Friboulet & D. Thomas, 'Systems biology: an interdisciplinary approach', *Biosens. Bioelectron.* 20, 2404–2407 (2005).

[93] H. Westerhoff & B. O. Palsson, 'The evolution of molecular biology into system biology', *Nat. Biotech.* 22, 1249–1252 (2004).

[94] D. Noble, 'Biophysics and systems biology', *Phil. Trans. R. Soc. A* 368, 1125–1139 (2010).

[95] H. Kacser & J. R. Small, 'How many phenotypes from one genotype? The case of prion diseases', *J. Theor. Biol.* 182, 209–218 (1996).

[96] W. Gibbs, 'The unseen genome: gems among the junk', *Sci. Am.* 289, 5 (2003).

[97] V. Lenin, *Materialism and Empirio-Criticism*, in Collected Works, vol. 14, trans. A. Fineberg, Progress Publishers, Moscow (1977(1909)) (our translation from Italian and http://www.marxists.org/archive/lenin/works/1908/mec/02.htm).

equivalence" cannot ensure that a normal tomato is the same as a transgenic one: if nothing else, in this case, tasting them would reveal the counterfeit – like pretending that a good copy of a Velasquez is identical to the original. It may deceive an inattentive observer but not an expert. The paradox is that Lenin, bitter critic of industrial capitalist civilization, did not realize that by this argument and by reducing the world to simple mechanics, he was proposing a science that mimicked the industrial process itself. In all seriousness, another figure claimed that "to understand was to build",[98] his idea of a world purged of awkward, irritating concern about Good, Beauty, Justice and Harmony, handed down to us by the Greek and Roman tradition, like so much other "useless stuff". This junk was blown away by determinism, but like all Pyrrhic victories, this too was short-lasting. Surprisingly, it was physics, the science that more than any other contributed to drawing new horizons, which undermined the foundations of the continuity and predictability of phenomena, depriving the scientific and philosophical concept of causality of any real meaning.[99] This it did with the Uncertainty Principle. Much later, medicine and biology are now travelling the same road. We are now witnessing a new Kuhnian revolution, this time in biomedical sciences.[100,101]

The process of progressive analytical dissection of the world, like anatomical dissection, is an operation carried out on cadavers. It enables us to acquire great power but not to understand.[102] A cadaver will never really tell us what life is. The problem is that only so much can be expected from this science: reduction of life to a sum of functions and machinery is not only an objective, but basically the greatest goal to which it can aspire.[103] Thus this science, and with it genetic engineering conceived as the last frontier, exhausts and completes Prometheus's hubris within itself, passing the columns of Hercules.

Genetic manipulation cuts the golden thread that joins us to the *Beginning* of creation. It is the latest advance of thoughtless science and demolishes the last fragile defence that protected us from the worst hubris: that expressed and conceptualized by the industrial world. In taking this step, it brings classical reason to an end. Though partial, this form of thought enabled us to understand without destroying, to conserve a sphere for science and to protect other areas of human action and life from science. Now scientific intervention, that seems to open unimaginable new horizons, forgets Maxwell's warning: when we interact with something to know it, we are already affecting it. The paradox was valid for subatomic physics, but can

---

[98] J. O. De La Mettrie, *L'Homme Machine*, Paris (1747).
[99] W. Heisemberg, *Der Teil und das Ganze* (the part and the whole), Piper & Verlag (1969).
[100] R. C. Strohman, 'The coming Kuhnian revolution in biology', *Nat. Biotech.* 15, 194–200 (1997).
[101] K. Strange, 'The end of "naïve reductionism": rise of systems biology or renaissance of physiology?', *Am. J. Physiol. Cell. Physiol.* 288, C968–C974 (2004).
[102] G. N. Amzallag, *La Raison Malmenée*, CNRS Edition, Paris (2002).
[103] G. Israel, *La Macchina Vivente*, Bollati Boringhieri, Torino (2004).

also be extended to other domains. Thus the genetic engineer does not realize that his intervention produces a cascade of reactions, most of them unpredictable, with such speed that it is impossible to establish exact cause–effect relationships. Before all the effects can be known, they generate new causes and so on to infinity. Result: the world we wanted to know and improve changes so much that we no longer know what changed or why and to whose benefit.

Shall We Conclude?

Some time ago, the Italian weekly magazine *Panorama* asked two famous geneticists their opinion on the safety of GMOs. Edoardo Boncinelli replied:

> The introduction of a new gene into the genome of a plant or animal is not *per se* associated with any risk. There is currently no scientific reason for fearing deleterious effects of transgenic foods on humans or the environment [though] it is clear that zero risk does not exist. [...] There is little rational in the arguments raised against transgenic organisms. They are scarecrows, symbols of the "sorcery of science" (or multinationals) and are raised by people who are against progress.[104]

We see that the reasoning and debate give way to insult in order to delegitimize the other party: those against GMOs are crazy obscurantists, enemies of progress and of humanity. Boncinelli does not seem to have read the report of the US National Research Council ... the criticism does not originate from the *Seattle people* or from some ideological preconception. It arose and developed within the scientific community that has no need of interpreters or committed spokesmen, a role that nobody seems to have delegated to Boncinelli. In the same issue of the magazine, the reply of Marcello Buiatti, a geneticist of world renown, was published:

> The basic scientific problem is the unpredictability of the modifications induced. It is one thing to let evolution try its genetic solutions and the environment select the fittest; it is quite another to insert a gene from another species, forcing genetic heritage and provoking an imbalance that cannot be predicted *a priori* [...] Among the characters modified in GM plants, two are particularly popular: resistance to insects and resistance to pesticides. In many cases the first effect of transformation was to harm the modified organism, making it unproductive. This is why multinationals concentrated on these modifications, using bacterial genes that did not have negative effects on the plants, in order to recoup investment in nearly 20 years of research. On their side they have patent law: any fragment of DNA isolated, useful for transforming an organism [...] can be monopolized for 20 years. This effectively prevents biotech industries from springing up in

---

[104] *Panorama*, 27 July 2000, p. 31 (our translation).

developing countries [...] Apart from the economic risks? Resistance to herbicides makes it possible to treat the growing crop with herbicides: productivity and herbicide use both increase, exposing the consumer to the risk of contamination [...] Moreover, genes for resistance to pesticides can be transmitted to other plants, altering ecosystems.[105]

It would be interesting to understand why two illustrious scientists have such different opinions on the same subject. The point I want to make is that this divergence of views, *within* the scientific community, does not facilitate understanding of the problem for normal people. The subject is complex and it is arrogant to claim to have all the answers. Thus scientists who deny the valid basis of mounting concern, considering precaution to be an *unprincipled position*,[106] would do well to humbly reconsider the impressive body of research and data that indicates that the road taken by biotech has non-negligible risks and unknowns, irrespective of any other ethical, philosophical or religious considerations. It is to be hoped that advances in knowledge will dissipate the many potential and real doubts and dangers.

It remains to be seen whether this is actually possible, especially in the short and middle period. In the meantime, it is legitimate to doubt and to demand the tests, controls and transparency that have so far been lacking. We discussed the question of labelling: no producer wants to indicate that its products contain material of transgenic origin. If they are not sure of their safety and "substantial equivalence", why deny this elementary right to consumers? Clearly someone is not telling the truth. Or, at least, not the whole truth.

As far as I am concerned, I wish to underline that I have never considered that "everything natural is good and everything artificial is bad". This has been spelled out in other circumstances, before the debate began:

The alleged safety of natural substances is based on a tacit misconception by which natural substances are artfully distinguished from synthetic substances, as if both were not chemical substances. The second assumption underlying this claim is that natural substances are by definition healthy and innocuous [...] 45% of natural chemical products tested are carcinogenic in at least one species [...] the world is full of chemical carcinogens and always has been.[107]

In the sphere of biotechnologies, the problem is *qualitatively* different. We are about to modify genetic heritage in an unpredictable and irreversible manner: we do not

---

[105] *Ibid.*

[106] H. I. Miller & G. Conko, 'Precaution without principle', *Nat. Biotechnol.* 19, 302–303 (2001).

[107] M. Bizzarri & A. Laganà, *Il Tramonto del Tumore*, U. Veronesi (Ed.), Errebian, Rome (1991), p. 176 (our translation).

know what awaits us "around the corner"; we know that much damage is possible. Perhaps nothing will happen, but the real point is: is it worth running the risk? And who is to take responsibility for doing so? Multinationals, governments, scientists? Are scientists completely sure they can remain "out of the scrum", especially when they are directly co-interested in the management of profits of biotech companies? This is not a polemical or disrespectful question to ask. Our reply to Meldolesi, busy defending the "objectivity of science", is not to be identified with the "objectivity of scientists", whose independence of judgement it is legitimate to doubt. Indeed, there is

> nothing new in the mingling of science and money: remember the research sponsored by the chemical industry that presumed to invalidate the DDT toxicity data published by Rachel Carson in *Silent Spring*? However in 1963 the scientific community rose up denying that research in specialist journals. Now scientific journals rarely publish letters of dissent. This may be because they are owned by a handful of publishers that often belong to financial holdings with investments in companies that advertise [...] The journals' rules on transparency are violated in 99% of cases [...] thus colleagues are invited to resist the power of multinationals [...] the price of economic success could be the integrity of science.[108]

These wise words of Sylvie Coyaud were published in the same journal as those of Meldolesi and should make us reflect, if only to understand why Pusztai was so suddenly fired after exposing the dangers of transgenic organisms. Meldolesi candidly recalls:

> [...] Though Rowett enjoys government grants, it received £140,000 from Monsanto. An embarrassing donation that according to many explains Rowett's behaviour towards Pusztai and its attempts to hush his version of the facts.[109]

In other words, re-phrasing a well-known statement of von Clausewitz, "food is too serious to be left in the hands of molecular biologists".

Let us all therefore take a step back and consider whether there are any alternatives, possibly less science-fiction-like, but safer, more environment-friendly and healthy. *Our* health and *our* environment.

> [...]The question which troubles the neutral observer is whether mankind's giant step is forward to a world in which there is less reliance on agro-chemicals, cheaper food and freedom from starvation, or backwards to the botanic and genetic equivalent

---

[108]S. Coyaud, 'Conflitti d'interesse: quanto è libera la ricerca scientifica?', *Le Scienze* 395, 110 (2001) (our translation).
[109]A. Meldolesi, *op. cit.*, p. 26 (our translation).

of BSE. To this conundrum there are no easy or quick answers [...] Others appear not to have learned the fundamental lesson of the BSE disaster which has so far cost around £4 billion and 35 lives: it is not enough to stand at the dispatch box or don a white coat and say "Trust me." It will never again be enough – especially in a world of giant corporations which have staked everything on pushing through the biotech revolution at maximum speed.[110]

Bionic Man

Opinion against GMOs is so strong because it addresses the same fears raised by human cloning. Even before one thinks about it, the problem elicits discomfort and exhumes distant memories of books by Meyrink, Shelley, Vogt and Asimov. Those pages were prophetic of monsters without souls or spirit, individuals produced in series in the incubators of a reproduction industry involving all three kingdoms – plants, animals, minerals – in disquieting mixtures. A second viewing of Fritz Lang's *Metropolis* can teach us more on bioethics than volumes of prolix outpourings. We can also seek consolation in Donald Duck, finding more wisdom in a comic strip than in the "science" pages of our newspapers. In a Gyro Gearloose story, the inventor, animated by a desire to build a machine that eliminates the bad side of people, inadvertently activates the device backwards on himself, resuscitating the joker that sleeps somewhere in all of us. The modified personality of the inventor causes one disaster after another, until the effect miraculously wears off. Brought before the judge, Gyro tries to defend himself:

> GYRO: I plead guilty, your Honour! But I assure you that I acted with the noble intention of improving people. Unfortunately the devil put in his tail, and instead of improving I got worse ...

> JUDGE: Defendant Gearloose, the Court condemns you to a fine of a thousand dollars [...] you must also swear never again to try to change human nature. *If people are the way they are, there must be a reason, even if we do not know it.*[111]

It may be an idea to send a free copy of the comic to all aspiring inventors, to remind them that "the road to hell is paved with good intentions" (medieval saying). Plenty escapes bioengineers who forget the warning of a father of modern biology:

> Equality is only meaningful because people are not the same [...] two aspects of equality [...] are important: the faculty to freely choose one's existence and assurance of a variety of

---

[110]Genetic haste, *The Guardian*, 17 February 1999.
[111] *Topolino* n. 385 del 13 aprile 1963, Mondadori, Milan (our translation).

> environments suitable for the genetic constitutions of different people.[112]

An assurance guaranteed (at least in words) by the constitution of half the world, but contradicted materially in the laboratory by the reconstruction of uniform, mechanized nature. The question we should ask is whether it still makes sense to ponder the identity and integrity of our nature. Have we not gone too far in the process of transforming our lives, that now depend so heavily on useless machines that bind us to a logic that has little to do with real human needs?

> [In other] technophiles and fashionable people, identification [with an artificial world] is so strong that these moral reactions, these fears about human integrity, make them laugh: for them clones mark the advent of the New Man of which they are the vanguard, the attainment of what they already are.[113]

As René Guénon would say, we are a parody: the new man that the metaphysical tradition of all civilizations writes about is a man who wholly incorporates his nature and transfigures it spiritually, becoming "what he is" (in the words of Nietsche). The new man foreshadowed by the modern world, fed with GMOs and *thought* by virtual reality, is actually an alien immersed in the infrahuman, dependent on extraneous logics that make him move like an element of a sophisticated software.[114] Have you seen the film *Matrix*? It is about real men kept in an incubator where they dream they are free, sentient and willing, while they are actually slaves of a computer programme that gives them a parody of existence instead of real life. The parody is certainly alluring: the virtual steak they eat is certainly more appetizing than the roots they eat when they seek refuge from the "machines" in the depths of the Earth, but it is still a virtual steak. This is the force driving the biotech industry: they will not sell us the tomato we know, but the *idea* of a tomato. We will need a lot of imagination to digest it ….

All this goes far beyond the dream of Prometheus. The concept of Prometheus arose from need: out of need of knowledge and power, Prometheus steals Zeus's fire. Humanity was cold and hungry and in the dark. But what need justifies GMOs? The silent revolution that is denaturing the foods that once graced our tables

> […] is not the answer to a need of society but was surreptitiously imposed on millions of consumers kept in ignorance by a handful of multinationals that perceived an opportunity to multiply their profits through genetic engineering.[115]

---

[112]T. Dobzhanzky, *Genetic Diversity and Human Equality*, Basic Books, New York (1973) (our translation from Italian).

[113]Encyclopédie des nuisances, *op. cit.*, pp. 37–38 (our translation from Italian).

[114]R. Guénon, *The Reign of Quantity & the Signs of theTimes*, 4th Revised Edition, Sophia Perennis (June 9, 2004).

Is this the freedom of science? Here, first of all, it is a question of *our* freedom!

The human condition was described incomparably by Pascal:

> When I consider the briefness of my life, devoured by past and future eternity, the tiny space I occupy and that I can see, submerged in the infinity of space I do not know and that does not know me, I am afraid, I am surprised to be here rather than elsewhere [...] Who put me where I am? By whose order and instruction was I assigned this place and this period? The eternal silence of infinite space terrifies me.[116]

Regarding this reflection of Pascal, the great biologist Theodosius Dobzhanzky observed that "space continues not to know us, but we begin to know something about it".[117] He asked whether "the silence of infinite space is more terrifying for us than for Pascal"?

Today the answer can only be *yes*: because if our knowledge has grown in a way that the French mathematician could never have imagined, the destructive potential of humans on nature has also increased. And when those spaces become less "infinite", perhaps there will no longer be humans to appreciate their beauty.

---

[115]A. Apoteker, *L'invasione del pesce-fragola*, Editori Riuniti, Rome (2000), p. 201 (our translation from the Italian).
[116]Cited in T. Dobzhanzky, *op. cit.*
[117]*Ibid.*

WITPRESS   ...*for scientists by scientists*

Environmental Impact

Edited by: **C.A. BREBBIA**, *Wessex Institute of Technology, UK;* **C.N. BROOKS**, *Greenfield Environmental Trust Group, USA and* **T-S CHON**, *Pusan National University, Korea*

Containing papers from the first International Conference on Environmental Impact and Development organised by the Wessex Institute of Technology, *Environmental Impact* fills the need for inter-disciplinary coverage of the most serious problems that affect sustainable development. The basic premise is that development projects need to consider the most pressing issues related to environmental impacts in order to provide complete solutions.

The current emphasis on sustainable development is a consequence of a general awareness of the various environmental problems that result from our modern society. This has led to the need to assess the impact of economic investments on the environment. Investment assessment and environmental economics need to be discussed in an integrated way, in accordance with the principles of sustainability. We must consider the social and environmental aspects of new investments, as well as possible environmental damage, including the destruction of natural resources and larger releases of waste and pollution into the environment.

The book covers such topics as: Environmental Policies and Planning; Environmental Assessments; Cost Benefit Analysis; Risk Analysis; Natural Resources Management; Rehabilitation Assessment; Social Issues and Stakeholders Participation; Decision Support Systems; Remediation Costs; Environmental Toxicology; Air, Water and Soil Contamination; Ecosystems Health; Environmental Health Risk; Biodegradation and Bioremediation; Ecotoxicity of Emerging Chemicals; Petroleum Contamination; Bioaccumulation and Biomonitoring; Brownfields Rehabilitation and Development; Monitoring of Brownfields.

WIT Transactions on Ecology and the Environment, Vol 162
ISBN: 978-1-84564-604-2   e-ISBN: 978-1-84564-605-9
Forthcoming 2012 / apx 300pp / apx £129.00

WITPRESS *...for scientists by scientists*

Food and Environment

The Quest for a Sustainable Future

Edited by: **V. POPOV** *and* **C.A. BREBBIA**, *Wessex Institute of Technology, UK*

The many advances in food production over the past century have made it possible to feed the whole of humanity. But food production and processing can have detrimental effects on the environment. Major challenges remain with industrial-scale farming. Higher productivity and larger volumes should not come at the expense of product quality or animal suffering.

Despite their importance, the consequences of food-related problems have not been sufficiently considered. It is essential to understand the impact that food production processes and the demands of rising living standards can have on the food consumed daily by the world's people. Of particular importance are the effects on human health and the well-being of the population, as well as the more general issues related to possible damage to the environment and ecology. This book includes contributions presented at the first international conference convened to examine these challenges.

Topics include: Impact of food production on the Environment; Contamination of Food; Food Processing Issues; Traceability and Temperature Control; Characterisation of Foodplants. The book will be of interest to food scientists and nutritionists, as well as agricultural, ecological, and environmental health experts interested in all these challenges.

WIT Transactions on Ecology and the Environment, Vol 152
ISBN: 978-1-84564-554-0   eISBN: 978-1-84564-555-7
Published 2011 / 256pp / £110.00

WITPress
Ashurst Lodge, Ashurst, Southampton,
SO40 7AA, UK.
Tel: 44 (0) 238 029 3223
Fax: 44 (0) 238 029 2853
E-Mail: witpress@witpress.com

WITPRESS   *...for scientists by scientists*

Sustainability Today

*Edited by: **C.A. BREBBIA**, Wessex Institute of Technology, UK*

This book contains additional research papers submitted for a meeting on sustainable development and planning organised in 2011 by the Wessex Institute of Technology (WIT). WIT has a long and very successful record of organising conferences on the topic of sustainability, which requires an interdisciplinary approach. Any sustainable solutions that are derived solely from the perspective of a single discipline may have unintended damaging consequences that create new problems. Thus effective sustainable solutions require the collaboration of scientists and engineers from various disciplines, as well as planners, architects, environmentalists, policy makers, and economists. These experts must not only communicate with each other effectively, but also understand the social aspects of the problem at hand. The contents of the book reflect that interdisciplinary approach.

ISBN: 978-1-84564-652-3   e-ISBN: 978-1-84564-653-0
Published 2011 / apx 370pp / apx £159.00

WIT Press is a major publisher of engineering research. The company prides itself on producing books by leading researchers and scientists at the cutting edge of their specialities, thus enabling readers to remain at the forefront of scientific developments. Our list presently includes monographs, edited volumes, books on disk, and software in areas such as: Acoustics, Advanced Computing, Architecture and Structures, Biomedicine, Boundary Elements, Earthquake Engineering, Environmental Engineering, Fluid Mechanics, Fracture Mechanics, Heat Transfer, Marine and Offshore Engineering and Transport Engineering.

WITPRESS   *...for scientists by scientists*

The Business of Biodiversity

M. EVERARD, University of the West of England, UK

The title of the book defines its primary focus: the linkage between biodiversity and business. The book is an accessible tome written primarily for people in business, but also in academia as well as in NGOs and in government, and those with a general interest. The text is deliberately brief to help the reader access the information quickly and is written in an engaging manner.

Biodiversity used to be an issue of high acknowledged importance, for example following the Convention on Biological Diversity (introduced at the Rio Summit in 1992) and the UK's first Biodiversity Strategy shortly thereafter. Yet, despite subsequent global commitments to reversing the rate of species loss made at the WSSD (Johannesburg 2002) and later endorsed by the EU (the Gothenburg Target), the key role of business in achieving a sustainable relationship with biodiversity has received little explicit attention over recent years. A re-appraisal has been due. Undertaking it has been rewarding since more recent insights into biodiversity suggest novel, self-beneficial responses for business.

ISBN: 978-1-84564-208-2   **eISBN: 978-1-84564-355-3**
Published 2009 / 208pp / £68.00

*All prices correct at time of going to press but
subject to change.
WIT Press books are available through your
bookseller or direct from the publisher.*

CPSIA information can be obtained at www.ICGtesting.com
Printed in the USA
BVOW022341070213

312705BV00004B/9/P